IMAGES
of America

MINING IN YUBA COUNTY

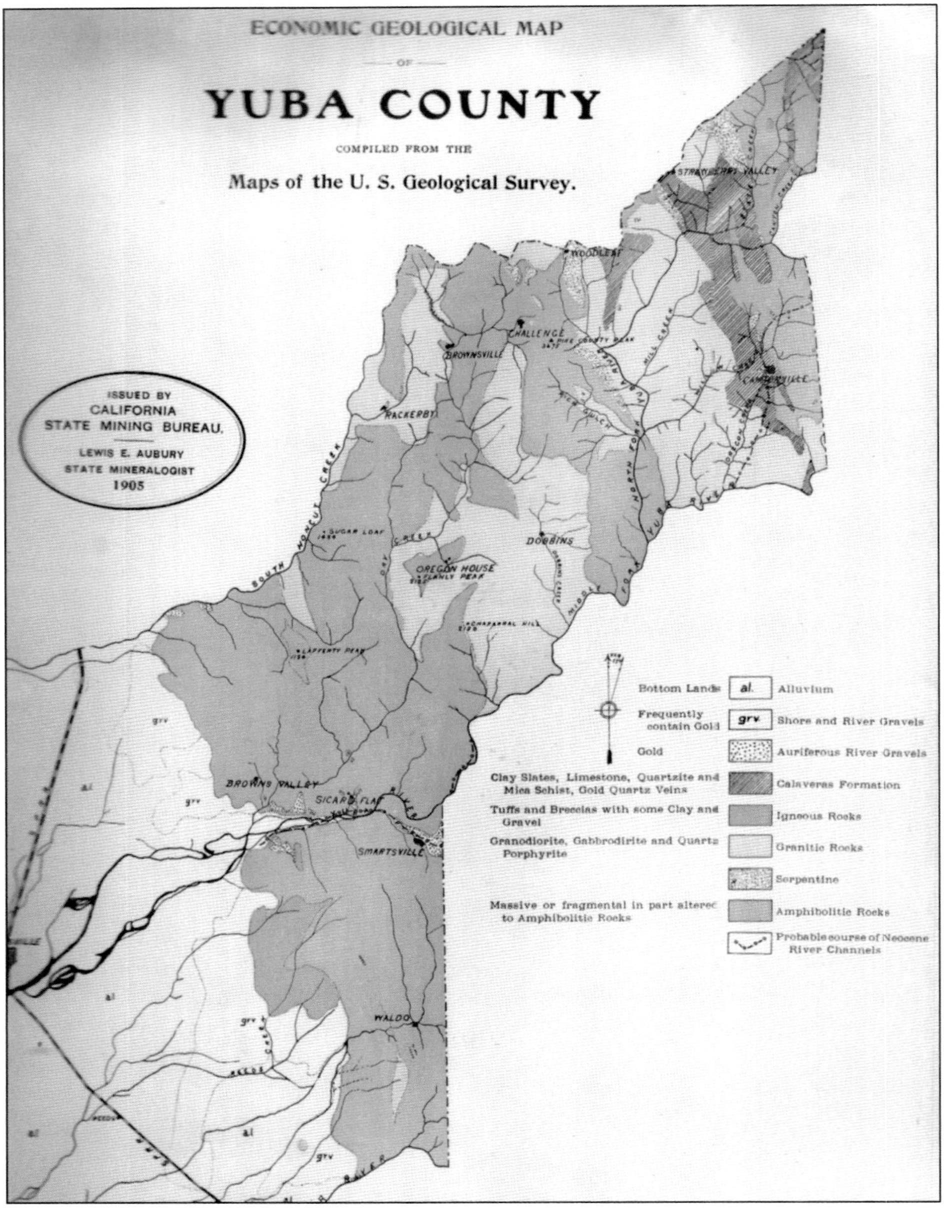

This 1905 map, issued by the California State Mining Bureau, shows the reason for the density of gold in Yuba County. The ancient Neocene river channels that run throughout the area contain auriferous river gravels that hold vast quantities of gold. The surrounding areas of shore and river gravel, and other outlying areas, frequently contain gold as well. Geologists know this area as the Smartsville Block, or Smartsville Complex. (Courtesy of the California Department of Conservation.)

ON THE COVER: This Risdon Iron Works dredge on the Yuba River is similar to the old style Risdon dredge produced by the Risdon Iron Works in San Francisco. As tailings from hydraulic mining filled the waterways, miners began dredging the stream to clear it and retrieve any valuable minerals not caught in the original sluicing process. This may have been the dredge used in 1910 by Mammoth Dredging Company. (Courtesy of the Marple/Hapgood Family.)

IMAGES
of America

MINING IN YUBA COUNTY

Kathleen Smith for YubaRoots

ARCADIA
PUBLISHING

ISBN 978-1-4671-3290-9

Published by Arcadia Publishing
Charleston, South Carolina

Printed in the United States of America

Library of Congress Control Number: 2015930920

For all general information, please contact Arcadia Publishing:
Telephone 843-853-2070
Fax 843-853-0044
E-mail sales@arcadiapublishing.com
For customer service and orders:
Toll-Free 1-888-313-2665

Visit us on the Internet at www.arcadiapublishing.com

*This book is dedicated to all who value history and its lessons—
allowing those lessons to guide us to a better future.*

CONTENTS

Acknowledgments

The people of Yuba County value the area's history immensely, and its historical community is very diverse. The commitment of these individuals and groups to preserving the stories of their past is truly remarkable, and many of them have graciously contributed images to this publication—something for which I am hugely grateful. Others have contributed by sharing their contacts and helping me find the information and images necessary to put this book together. Some of them are, in no particular order: Sue Cejner-Moyers, who tirelessly teaches history to children through storytelling and historical character presentations; Gina Zurakowski, the librarian at the Local History Archives of Yuba County Library; Chris Ward, archaeologist and tour guide extraordinaire; Doug Criddle and Ruth Criddle, who preserve the memories of the lost dredger towns of Hammonton and Marygold and keep former residents in touch by holding annual reunions; Rosemarie Mossinger and the Yuba Feather Museum; and Ruth Mikkelsen and Roberta Sperbeck D'Arcy of Brown's Valley. Special thanks are also owed to my fellow SCRFI's (members of Smartsville Church Restoration Fund Inc.), who have also been bitten by the preservation bug. Kit and Janet Burton, who encourage me, and Brian Bisnett, for preserving former mining lands that exhibit the vestiges of mining in Yuba County for wildlife habitats and recreation sites that future generations will have the opportunity to experience for themselves. Thanks also to Kathy Sedler, the founder of YubaRoots, for devoting countless hours to building and maintaining the invaluable YubaRoots website. The information found on that site inspired and aided me in following my own personal genealogical journey, as I know it has countless others. When Kathy dropped out of this project, I continued on to show my appreciation of her efforts and YubaRoots. Gratitude is also owed to history professor emeritus David Rubiales, who generously offered to review the manuscript and "keep me honest." I would also like to thank Jeff Ruetsche at Arcadia Publishing, for getting the project back on track and keeping it moving through to completion. Finally, I am grateful to my family for humoring me in my zealous quest to mine the gold nuggets of history—especially my mom, who continues to make it possible for me to do so.

Introduction

Imagine a serene river valley, leading up to rolling foothills and fed by pristine tributaries that begin in the snow-capped Sierra Nevada Mountains. Concealed in the foothills and the streams is a treasure, formed there 65 million years ago. The river and high cliffs begin to slowly reveal this hidden treasure—the water washing and wearing away parts of the earth until shining nuggets escape into the streams. The native people living in the area know about these trinkets but don't have a purpose for them or place value on them. It is not until settlers come to this place and recognize the nuggets as gold that the tranquility of the valley is disturbed by a mass migration of treasure hunters seeking out their share. Thus began the California Gold Rush of 1849.

The history of mining in Yuba County is the story of that mass migration, and of the inevitable way in which the development of the mining industry intertwined with the evolution of the county, the little towns that grew there, and the people who called both of these worlds home. California was admitted into the union on September 9, 1850, and Yuba County was one of the state's original 27 counties. Founded on February 18, 1850, it originally included parts of Nevada and Placer Counties.

There was no Yuba County in 1849, but there was gold in the rivers and the hills—a lot of it. Countless people migrated to the area—traveling from across the United States and around the world to try and make their pile. Some came overland, following the route taken by the infamous Donner Party, or by the Oregon route. Others traveled by ship and sailed around Cape Horn. More still combined ship travel with an overland crossing—disembarking at the isthmus of Panama and crossing it to board another ship going up the Pacific coast. No matter how they came, it was a hard journey, and one filled with risk. Some people made their fortunes, but many went away sorely disappointed. Regardless of the odds, there were always more travelers flooding in to participate in the great gold rush.

They say that nothing exists in a vacuum, and so it was with Yuba. The events of the gold rush created a ripple effect, and even though a bid for statehood was in the works, it is thought that the discovery of gold in California hastened California's admission. There is also speculation that events at Rose Bar, on the Yuba River, led to California being admitted as a free state. It has been further said that California's gold, deposited into the United States Treasury, gave the Union the funds it needed to win the Civil War.

When considering California's current events, it is easy to recognize that many of today's issues and concerns are strikingly similar, or in some way related to the events of the gold rush. Issues such as water rights, mineral rights, flood control, land use, maintaining habitat for migrating water fowl, and restoring spawning grounds for salmon in our rivers all have their roots in the gold rush.

As of this writing, Northern California is going through one of its periodic droughts—something that happened in the past much as it does today. In 1870, water wasn't plentiful enough to work the mines, but there were subsequent years when rain was too plentiful and flooding occurred—

instituting flood control was a major issue. In the more than 100 years since hydraulic mining stopped, the effects of the industry's tailings are still a factor in flood control. Even now, floods can be attributed, at least in part, to the industry, and a concern for the strength of the levee system is at the forefront of the government's agenda.

Further, rivers reclaimed from the tailings and made navigable have since filled with silt again, and any effective method of clearing them must be weighed against the damage that could be done to local wildlife. The water issues currently being decided in the state legislature, and the idea of building giant tunnels down the Central Valley to send water to the southern part of the state, are hotly contested.

It is not grandiose to say that the gold rush and the mining industry are more responsible for making Yuba County what it is today than any other historical event. It is what put Yuba County on the map, the thing that inspired innovations across multiple industrial fields around the state and the country, and it has made Yuba the home of many people who appreciate history and learn from it—hoping to make this part of California golden in a new and positive way.

At present, there are many exciting efforts to return Yuba County to the prosperity of the gold rush—using local resources wisely and, at the same time, making it a beautiful place to live and visit. There are many family farms and small businesses here, growing and producing healthy foods. Fruits and nuts are grown, packaged, and sent all over the world, and locally grown rice is renowned worldwide for its superior quality. Naturally raised, grass-fed beef and lamb has been in style here since the days prior to the gold rush—when the land was used almost exclusively for grazing—and it is now growing in popularity with consumers around the country. Many artisan products, such as olive oils, jams, and honey, have been revived; extensive vineyards have been planted; and award-winning wineries abound. Historical and cultural organizations are doing restoration and preservation work, while educational activities and local land trusts are preserving wildlife habitats—honoring nature and creating spectacular recreational opportunities for everyone. Historical, recreational, and agricultural tourism may just be the new gold rush of Yuba.

One

Every Man for Himself

Rules of Conduct and Organization

On June 2, 1848, five months after gold was discovered at Coloma, Jonas Spect found gold on the Yuba River—just below what is now known as Rose's Bar. Due to its particular geological formation, the extent of Yuba's mineral richness was previously unknown. Spect's discovery began a rush to the Yuba River, and soon, industrious individuals swarmed the banks of the Yuba in search of gold. The tributaries were covered with prospectors, while still more swarmed the ravines. Up to this point, the land around the Yuba River had only been occupied by some native groups and a few rancheros with Mexican land grants. These residents already knew of the area's abundant natural resources, like ample fresh water, plentiful fish and game, grass for feed, and temperate climate. There was also plenty of timber for fuel, buildings, and tools.

In that tentative time, prior to California's statehood, the grasp of Mexican rule had been loosened and there was no prevailing law in place. It was every man for himself until the prospectors took matters into their own hands. On April 10, 1849, the miners at Rose Bar held a meeting to establish laws and regulations by which their mining enterprises should be governed. Jonas Spect presented a set of mining laws, which were adopted and were claimed as the first mining laws framed in the state. In June 1850, it became apparent that more regulations were needed; delegates from each mining camp were chosen, and the Miners Court was created. Now, on a broader scale, they drew up a set of guidelines by which to govern themselves. Interestingly, these guidelines included the decision to not grant a claim to anyone using slave labor. The independent spirit of the miners was demonstrated by making sure that the opportunities of individuals were not infringed upon, and with the establishment of the court, they essentially became the first lawmakers in Yuba.

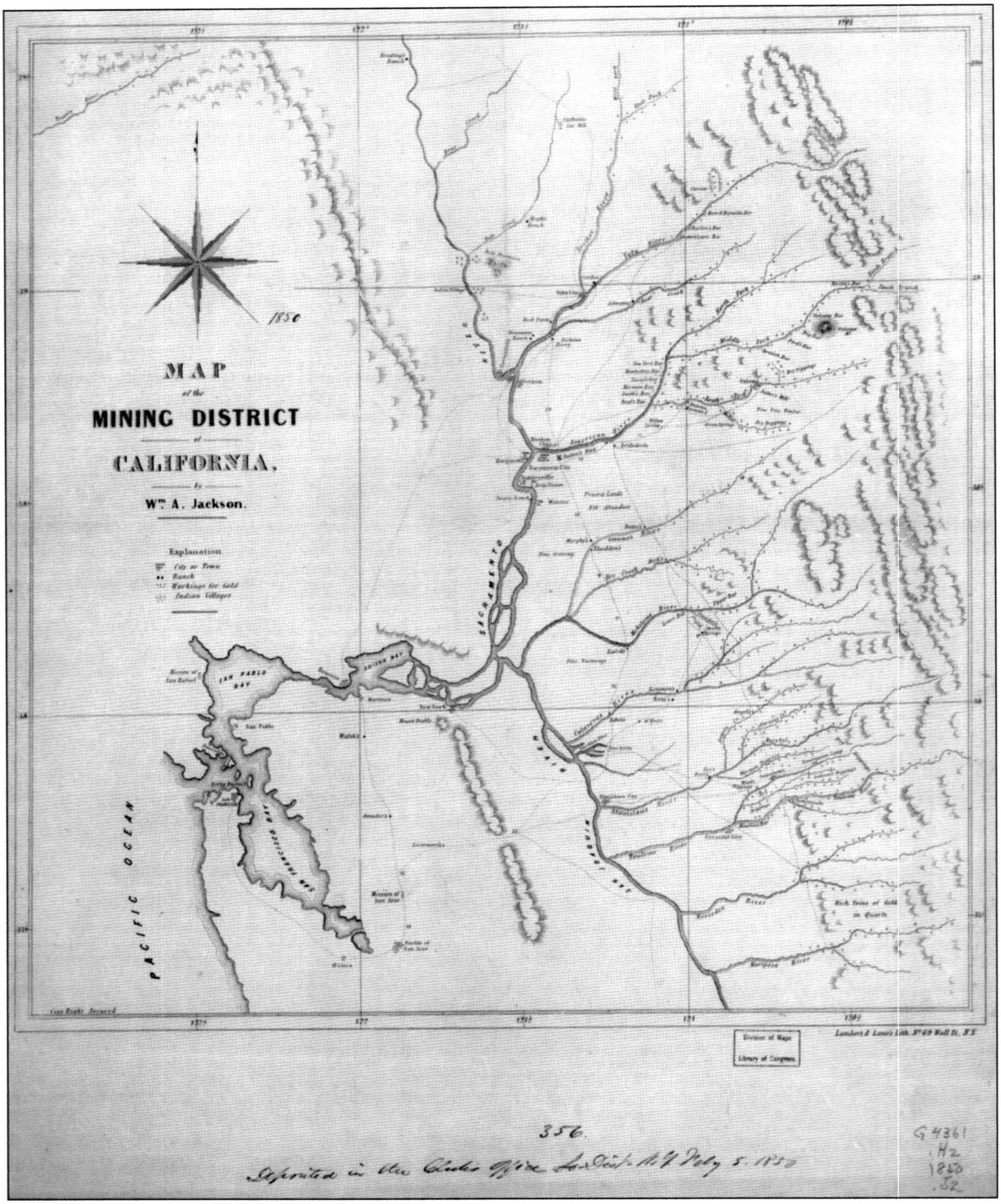

This 1850 map of the mining district of California compiled by William A. Jackson shows towns, ranches, mining works, and Indian villages. Traveling to Yuba was accomplished in large part over water; from the San Francisco Bay through the San Pablo and Suisun Bays, then up the Sacramento River to Sacramento, and finally to the Yuba River region. Only a few locations in the Yuba vicinity are listed, including Johnson's Rancho on Bear Creek, which was a transit point for parties to and from the East; Cordua's Ranch, where stock was raised and there was a trading post; several mining settlements; and numerous Indian villages. Yuba City is on the map but not Marysville. Rose and Reynolds' Bar later became Rose's Bar. (Courtesy of Library of Congress.)

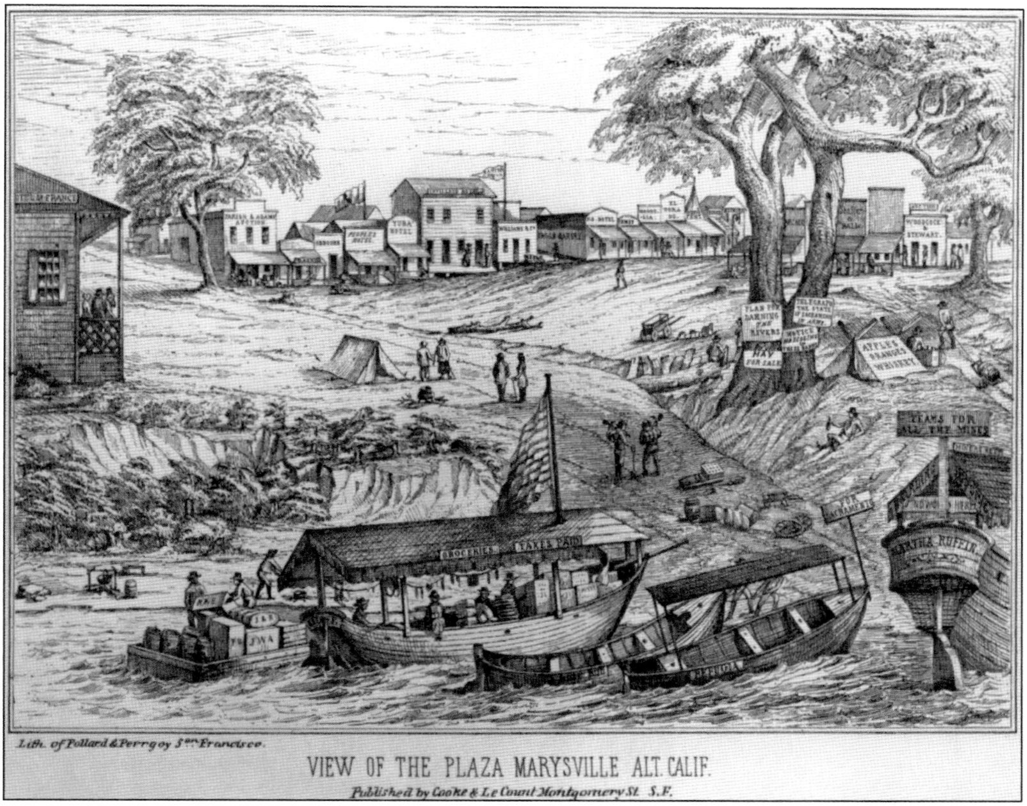

Marysville was a town born of the Yuba mining boom. In December 1849, a town was laid out and called Yubaville. It was believed that the location would be the perfect hub for the transportation of people and supplies to the mines. It was later renamed Marysville in honor of Mary Murphy Johnson Covillaud. It also eventually became the seat of Yuba County. (Courtesy of Library of Congress.)

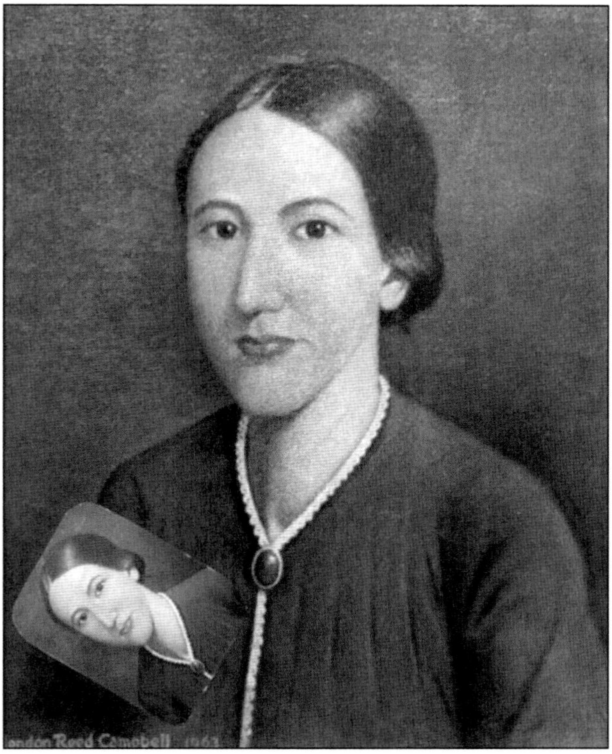

Mary Murphy Johnson Covillaud was a survivor of the ill-fated Donner Party and the wife of Charles Covillaud, one of the founding members of Marysville. (Courtesy of the Yuba County Library.)

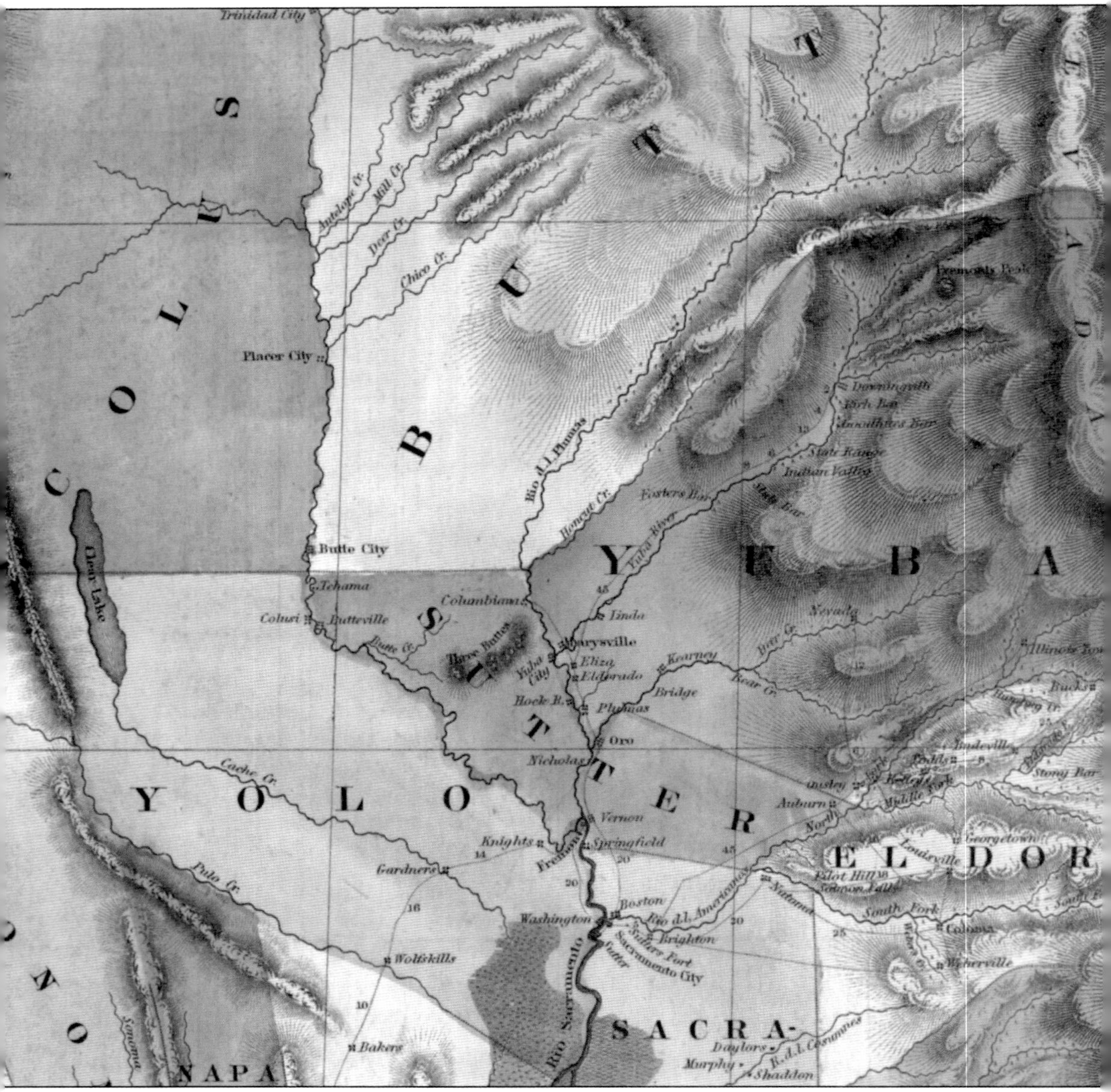

This detail from the 1851 *Map of the Gold Region in California* by Charles Drayton Gibbes shows the area originally covered by Yuba County. One of California's original 27 counties, Yuba was founded on February 18, 1850, and included parts of Nevada and Placer Counties. The county was divided and reconfigured several times before arriving at its current configuration. (Courtesy of Library of Congress.)

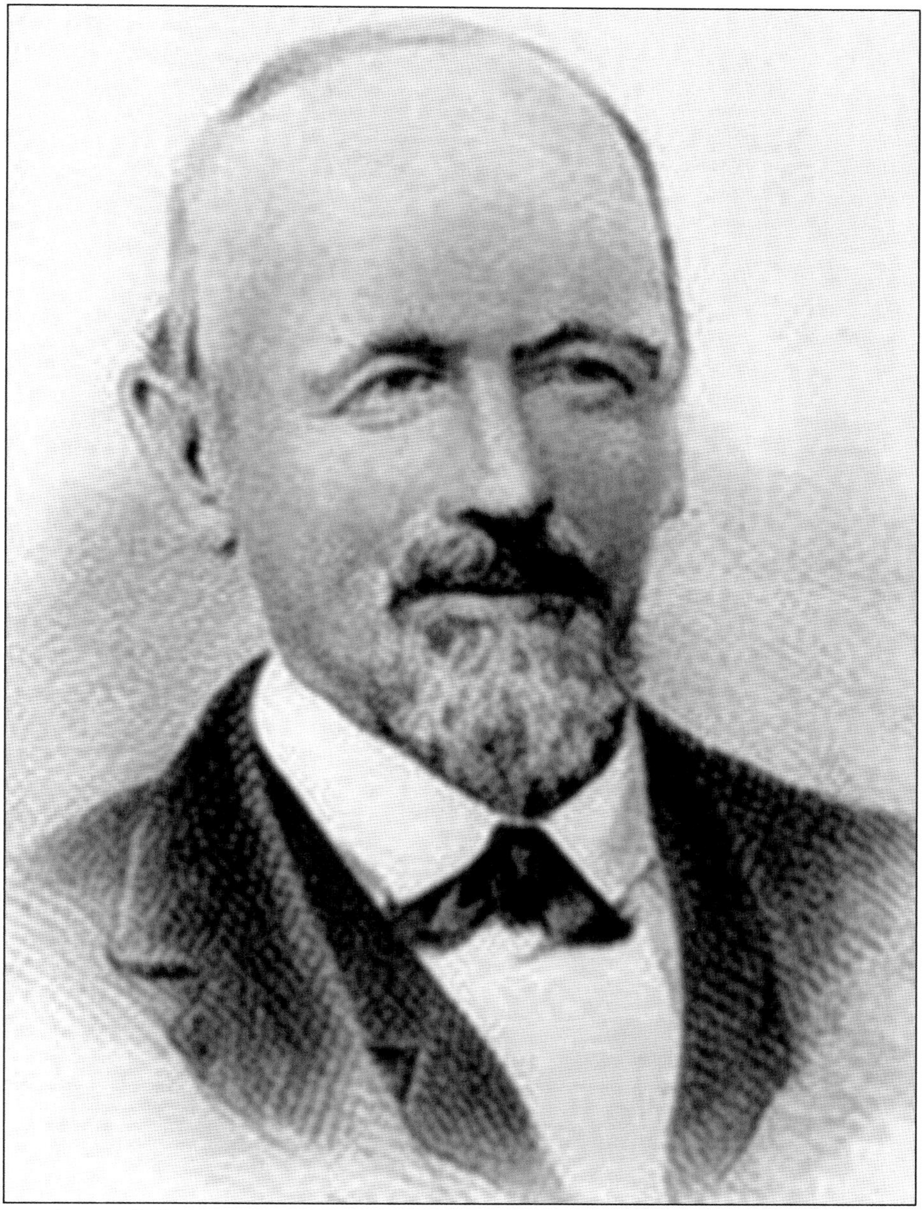

In December 1847, Jonas Spect was in San Francisco, but by the spring of 1848, he was waiting for a party to gather at Johnson's Ranch so he could return East. Due to the excitement of the discovery of gold, they were unable to gather enough people for the trip and Spect decided to go with some native guides, hoping to determine if there were paying quantities of gold in Yuba. The first several prospecting sites yielded nothing, but just as he was about to quit, on June 2, 1848, Spect found several good-sized nuggets. He continued prospecting until November, when he decided to pursue other ventures. He established a store at the confluence of the Bear and Yuba Rivers. In March 1849, he began operation of a ferry that crossed the Sacramento River at Vernon, established the town of Freemont on the Yolo side, and opened a store and a hotel. Spect was elected as a delegate to California's Constitutional Convention and later served as a state senator from Sonoma. He was living in Colusa at the time of his death. (Courtesy of Sacramento History Collection.)

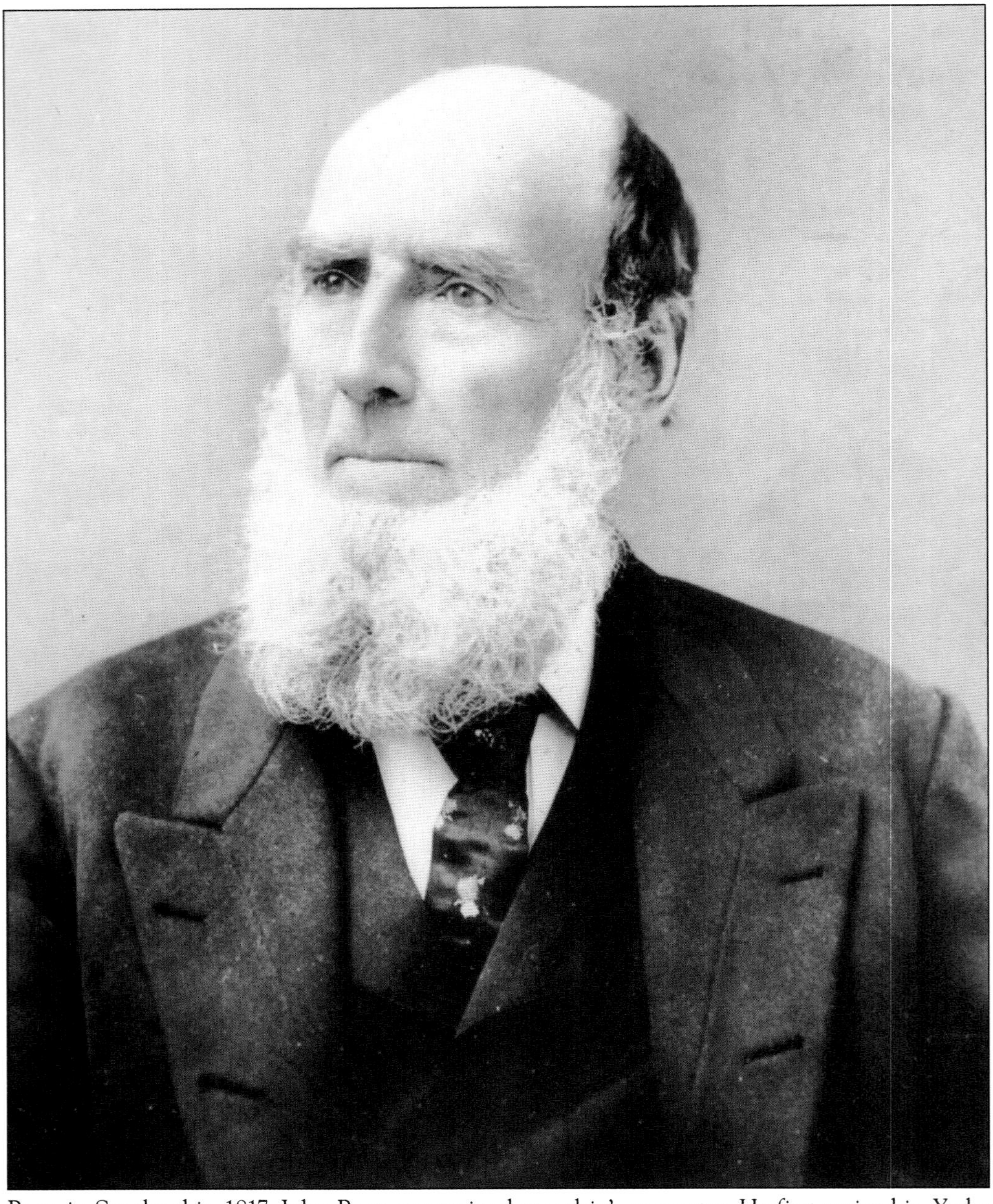

Born in Scotland in 1817, John Rose was trained as a ship's carpenter. He first arrived in Yerba Buena in 1839 but didn't settle there until 1844. He was the first treasurer of that new town and was granted a town lot. In a partnership with Chino Reynolds, he was contracted to build buildings and ships. News of the discovery of gold came while Rose was building a mill for Salvatore Vallejo in Sonoma. His crew wanted to leave immediately for the American River, but Rose was able to convince them to finish the job with the promise that he would take them there. When they arrived at Sacramento, Rose heard about gold found on the Yuba and promptly filled his wagon with supplies, headed out, and established his claim. Rose and Reynolds built a trading post at this location, which was named for them. The original intent was to trade with the natives and supply the miners. Rose also dabbled in other businesses, like operating a ferry, and real estate, by laying out the town of Linda. (Courtesy of Yuba County Library.)

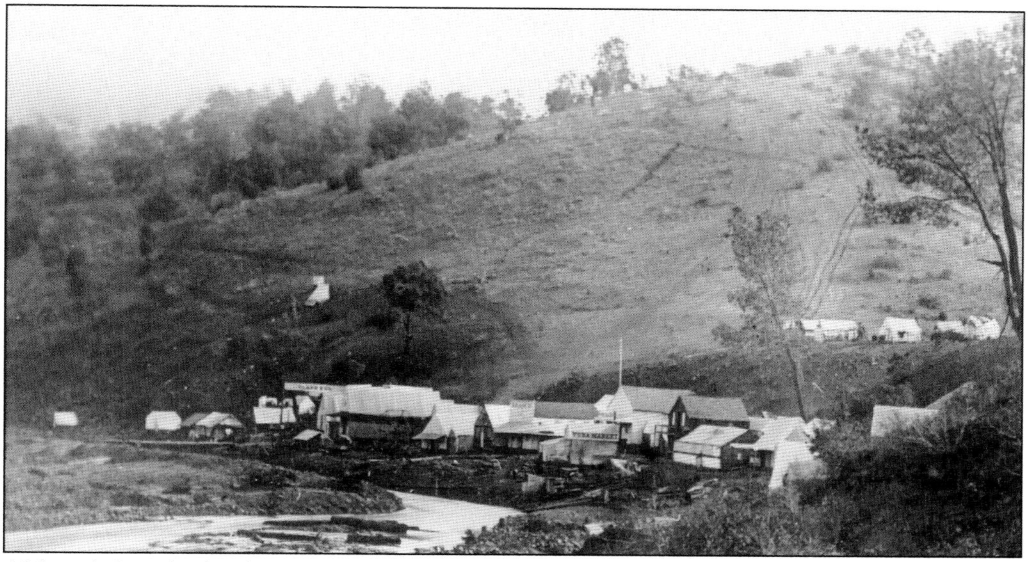

Although Rose had a claim, it was not his primary business. Reportedly, he paid natives to shovel dirt into sluice boxes in exchange for beef. As seen above, Rose's Bar had grown significantly by 1852. Many stores, saloons, and hotels had sprung up to serve the miners; even a Masonic lodge and a church can be seen. Rose and Reynolds took on a third partner, George Kinlock, and procured land to raise stock. Below is a map of the area covered by Rose's rancho. The business was not profitable, so the partnership dissolved. In order to make a living, Rose was once again contracted to build buildings, which he did throughout the county. He eventually settled on his ranch, married, and raised a family. (Above, courtesy of Bill Fuller; below, courtesy of Yuba County Library.)

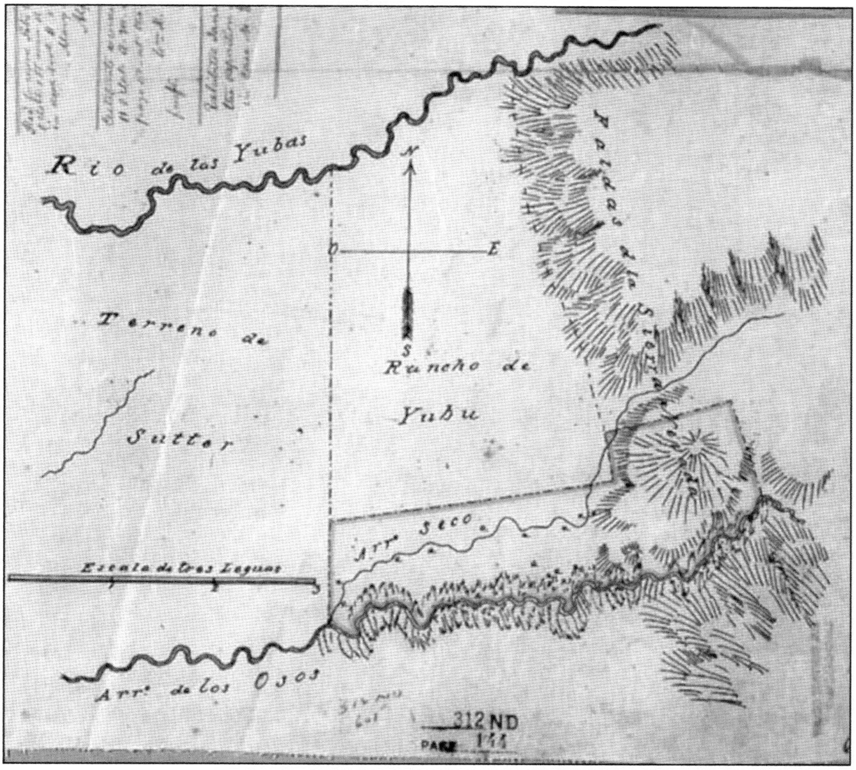

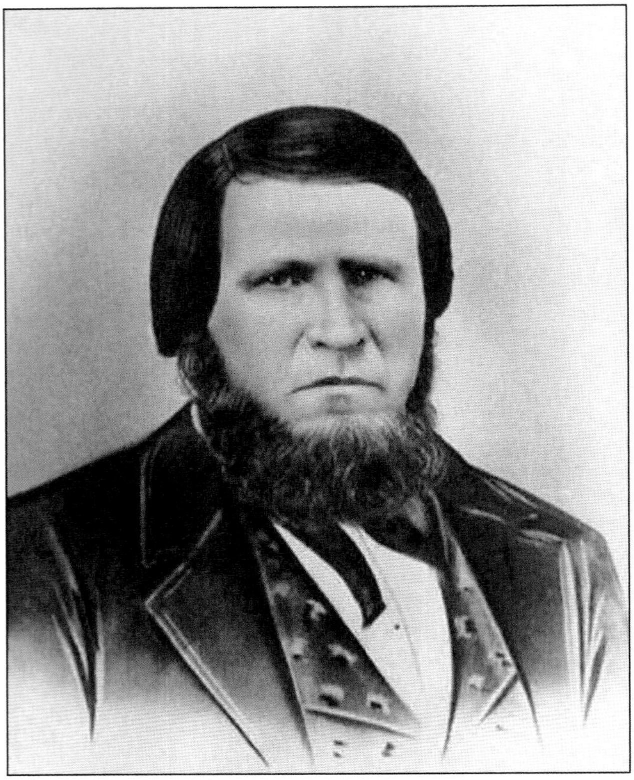

David Parks had been a farmer in Ohio and Indiana. In 1948, while traveling to Oregon with his family, he heard of the discovery of gold in Yuba and changed course. He and his family arrived at a bar that was eventually named in their honor. He opened a store and mined, but only stayed in the area for a few years, leaving the bar—which had become a thriving camp—in 1852 to return to farming. (Courtesy of Yuba County Library.)

In June 1848, Park's Bar—located just south of Rose's Bar—hosted a company of miners who arrived, began working at the bar, and soon moved on—thinking it paid too little. Others arrived shortly after, and by 1849, a thriving town had sprung up. Here, miners shovel dirt into their sluice boxes. The town is visible in the background. (Courtesy of the California State Library.)

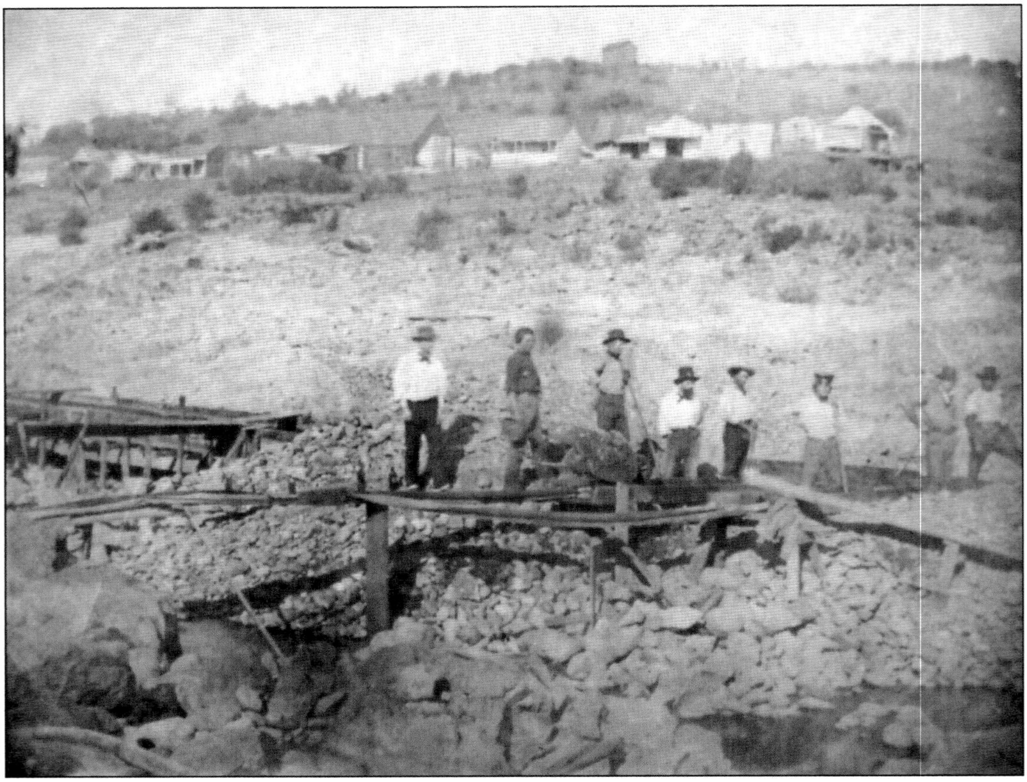

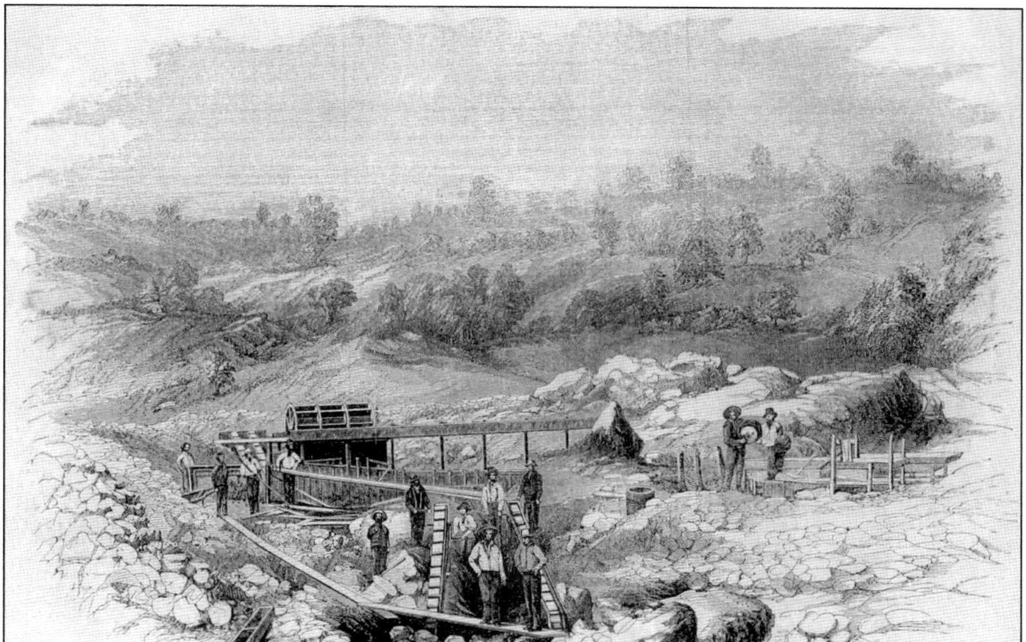

NOTION COMPANY'S WORKS, CALIFORNIA.

The March 25, 1854, issue of *Gleason's Pictorial* ran a two page spread entitled "Scenes in California," which included these drawings of scenes that played out on the Yuba River. The above engraving shows the Norton Company's works at Barton's Bar, and the one below is described as "a view of the Parks' Bar Company's Works showing the flume, men at work, pumps, wheels, etc." (Both, author's collection.)

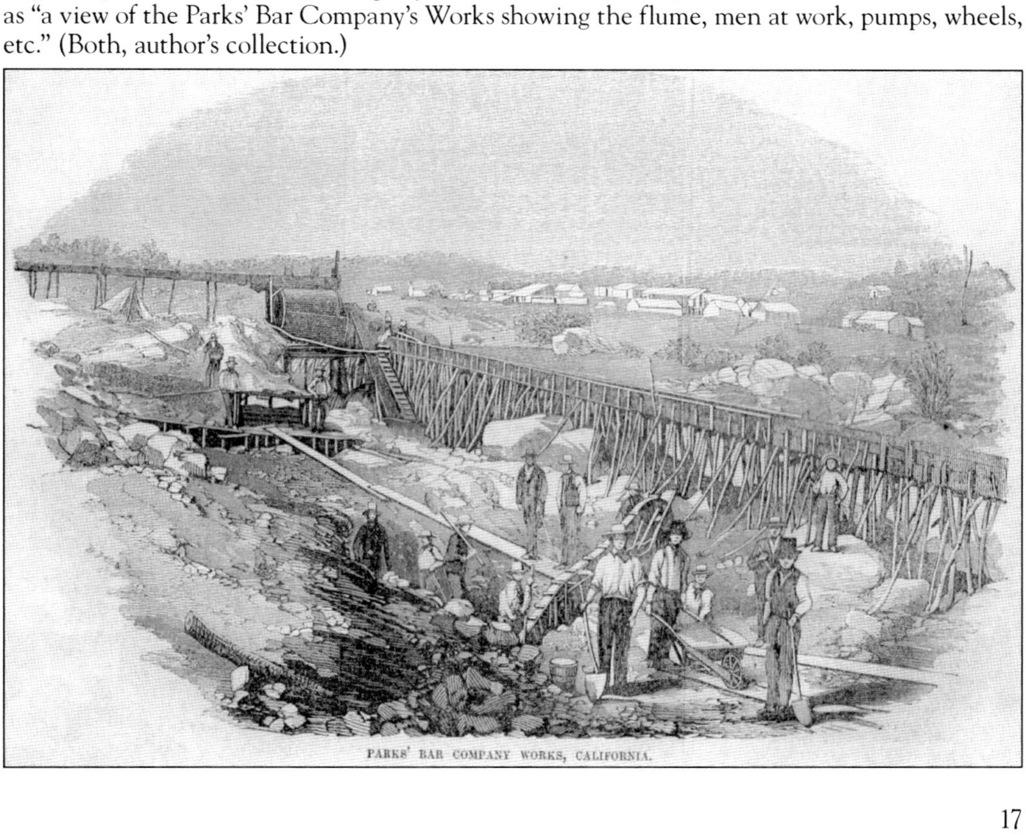

PARKS' BAR COMPANY WORKS, CALIFORNIA.

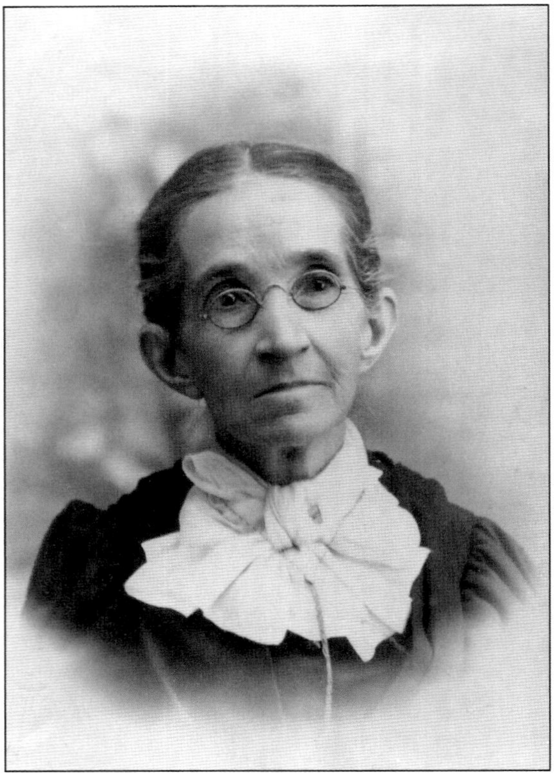

William McFadden Foster and his wife, Sarah Ann Charlotte Murphy Foster, survived the Donner Party incident of 1846 as they traveled to California. In 1849, they moved to Yuba County. William Foster was one of the founders of Marysville and also spent some time mining at the bar situated on the west bank of the North Yuba, between Willow and Mill Creeks. William opened a store, and a well-developed town grew around it. Foster's Bar and Foster Bar Township were both named for him. The Fosters eventually left the area and did not return. (Both, courtesy of Yuba County Library.)

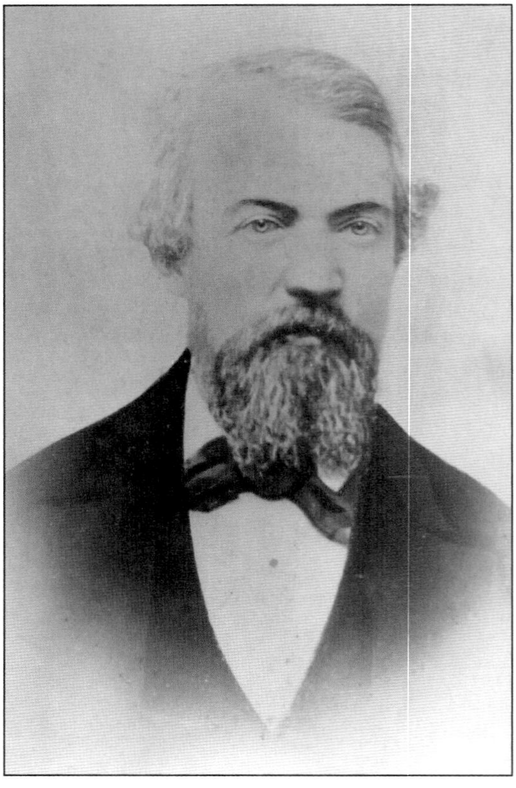

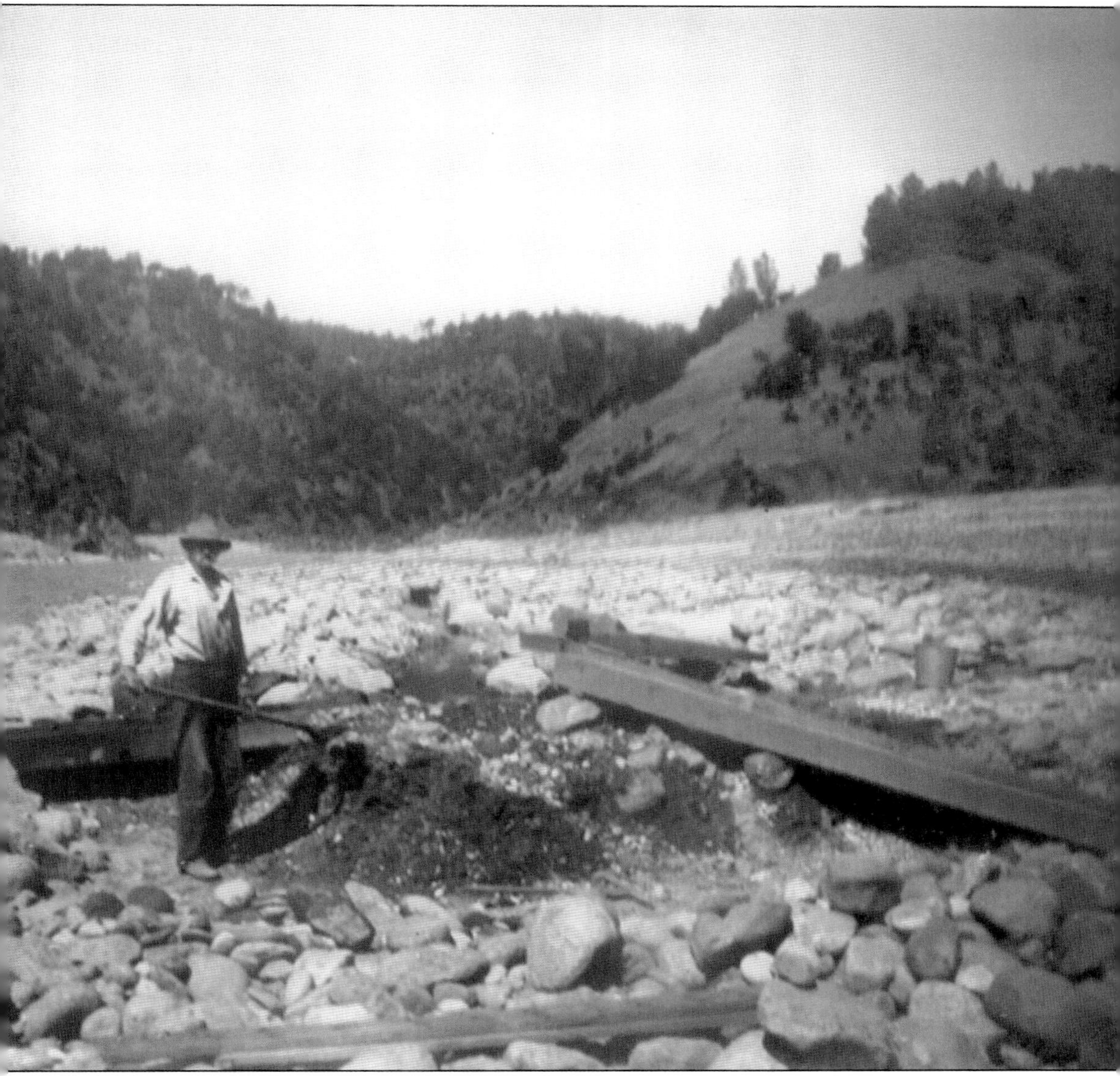

In 1850, the population of Foster's Bar ranged between 500 and 1,200 people, depending on the season. Hotels and saloons sprang up, and whiskey sales did a brisk business. That year, the miners also created a local government and regulated claim sizes. Following the flood of 1862, the bed of the river rose 15–18 feet due to tailings from the upper mines, burying many claims under the debris. (Courtesy of Mollie Plitzko.)

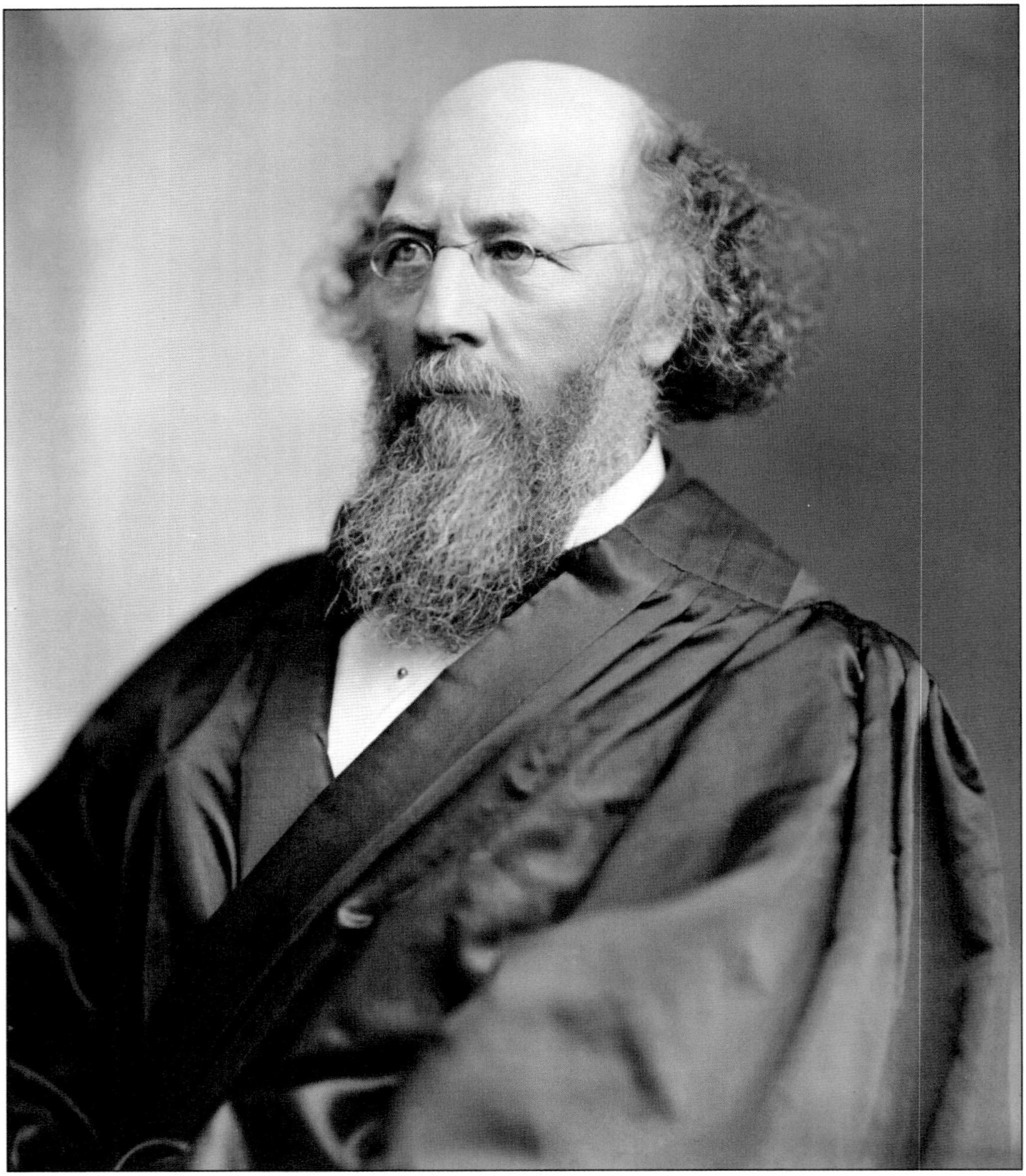

Stephen Johnson Field arrived in Marysville in January 1850, seeking work as an attorney. He served as the new town's first mayor, and in 1851, as a legislator in the State Assembly—representing Yuba County and the new Sierra and Nevada Counties. He also passed into law the miners' guidelines for self-governance in relation to mining claims and regulations, which was later adopted by the US Congress. (Courtesy of Library of Congress.)

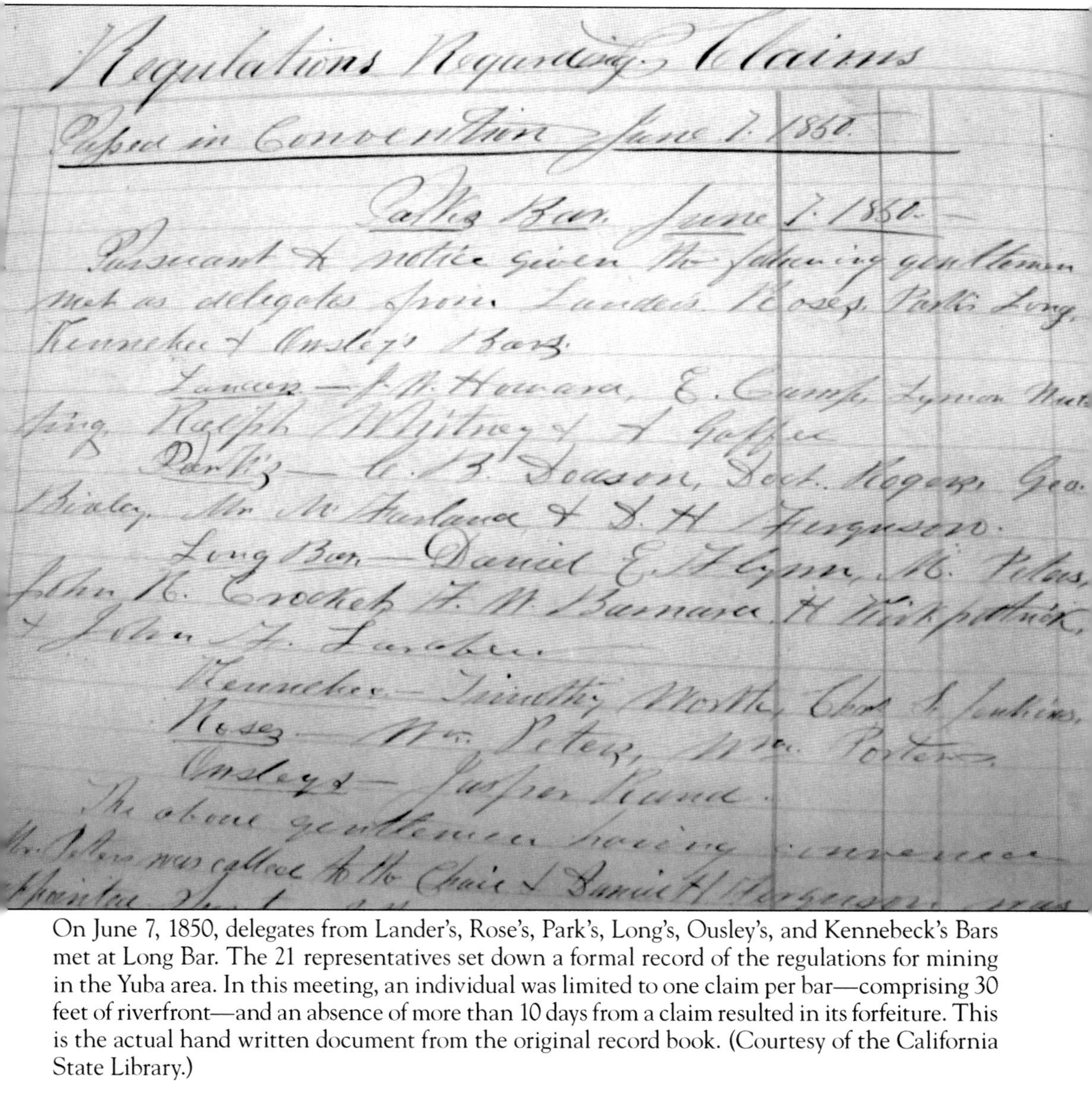
Regulations Regarding Claims
Passed in Convention June 7. 1850.
Long Bar. June 7. 1850.
Pursuant to notice given the following gentlemen met as delegates from Landers, Roses, Parks, Long, Kennebec & Ousley's Bars.
Landers — J. H. Howard, E. Camp, Lyman Nutting, Ralph Whitney & A. Goffee
Parks — C. B. Dawson, Doct. Rogers, Geo. Binley, Mr McFarland & D. H. Ferguson.
Long Bar — Daniel E. Flynn, M. Pilas, John R. Crockett, F. M. Barnard, H. Kirkpatrick, & John F. Lander.
Kennebec — Timothy Worth, Chas. S. Jenkins.
Roses — Mr. Peters, Mr. Porter.
Ousleys — Jasper Reed.
The above gentlemen having convened
Mr. Pilas was called to the Chair & Daniel H Ferguson was
appointed

On June 7, 1850, delegates from Lander's, Rose's, Park's, Long's, Ousley's, and Kennebeck's Bars met at Long Bar. The 21 representatives set down a formal record of the regulations for mining in the Yuba area. In this meeting, an individual was limited to one claim per bar—comprising 30 feet of riverfront—and an absence of more than 10 days from a claim resulted in its forfeiture. This is the actual hand written document from the original record book. (Courtesy of the California State Library.)

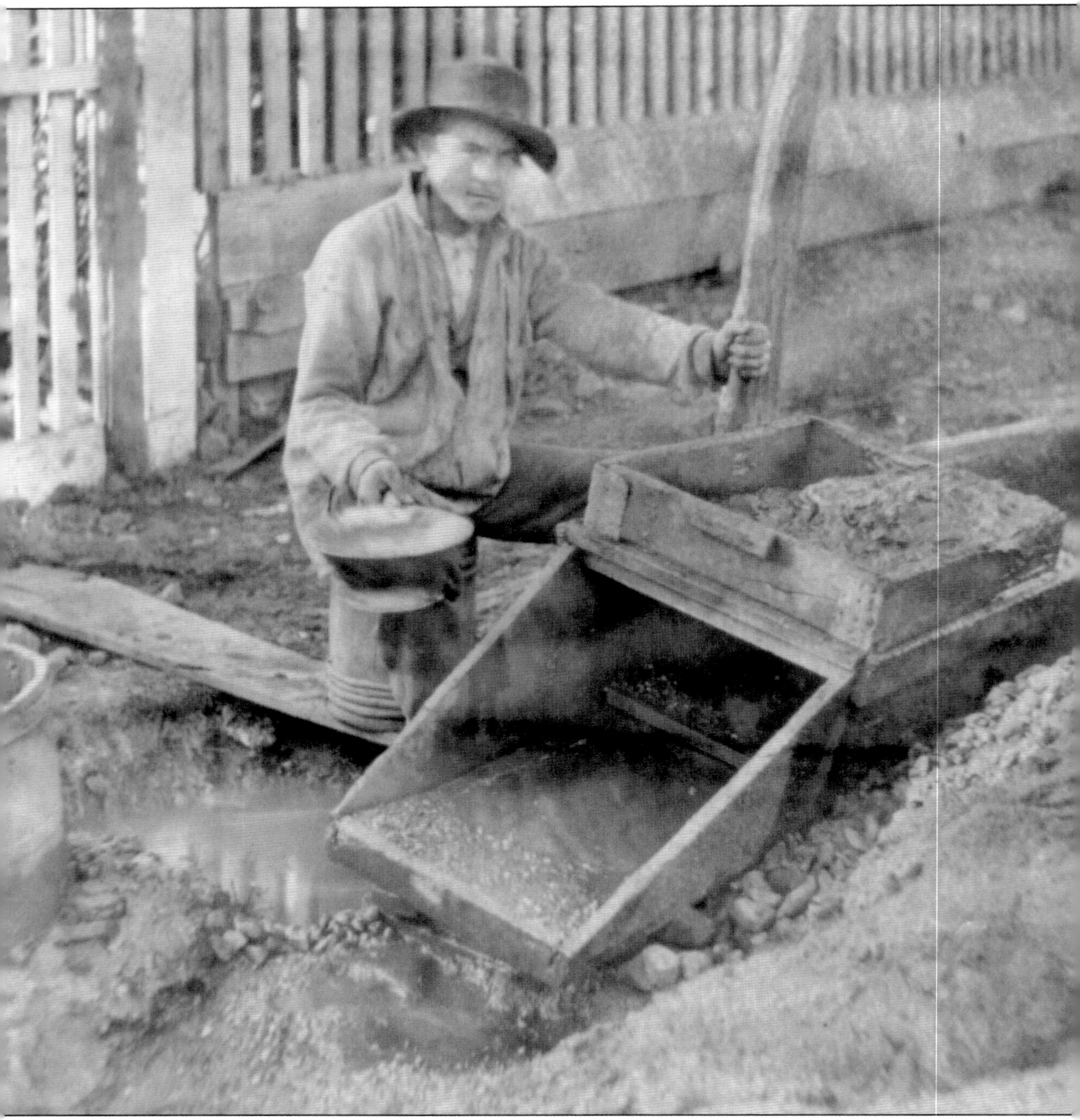

Placer gold is gold that has been freed from its host rock. It can be found in gravel deposits or in rivers and streams. Panning was the earliest method to separate placer gold from the sand and gravel in which it was contained. The pan was filled with gravel and water, and the miner gently swirled and shook it—allowing the heavier gold to settle, and the water to slosh out the lighter debris. Not much gravel could be worked this way, so other methods were devised. The rocker in this photograph is a rectangular wooden box mounted on a rocking mechanism. A sieve is at the top, and at the bottom are a series of riffles. The dirt and rock were dumped into the top, followed by a bucket of water. As it was rocked by hand, the larger rocks dumped out of the sieve, and as the water worked its way through, it washed the dirt and sand out. The heavy gold fell to the bottom and was caught in the riffles. (Courtesy of Yuba County Library.)

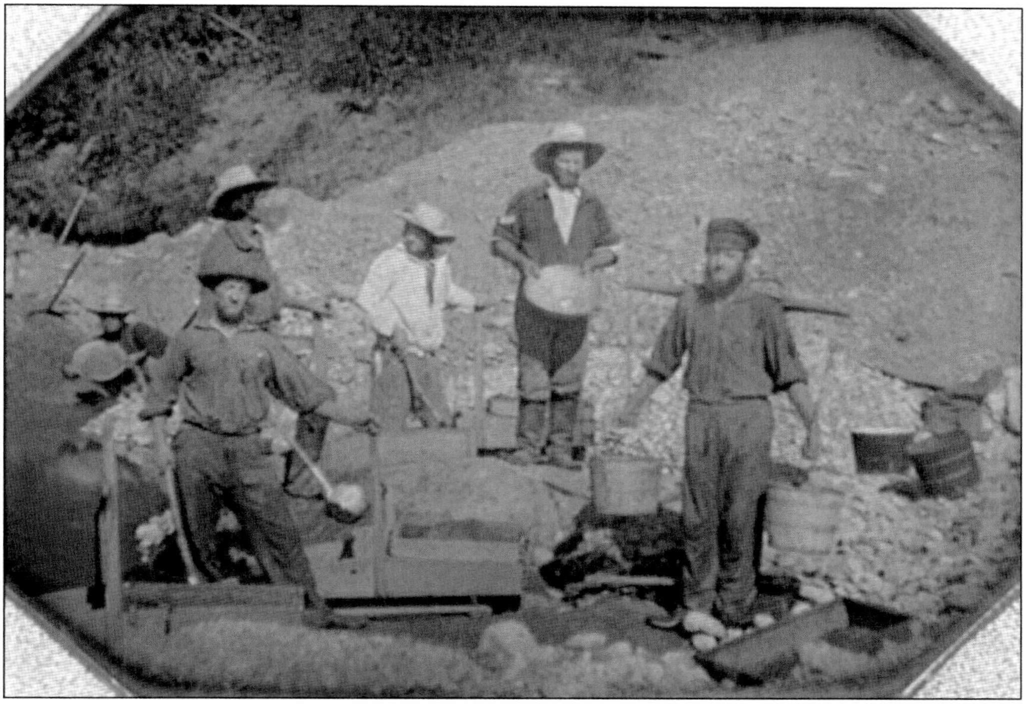

The Long-Tom was a little larger than the rocker and could handle a little more volume. It could also be used by more than one person at a time. Gravel was shoveled in and water was run through it, washing away the larger, lighter matter and leaving the heavier gold to fall to the bottom of the trough and be trapped in the riffles. (Courtesy of Yuba County Library.)

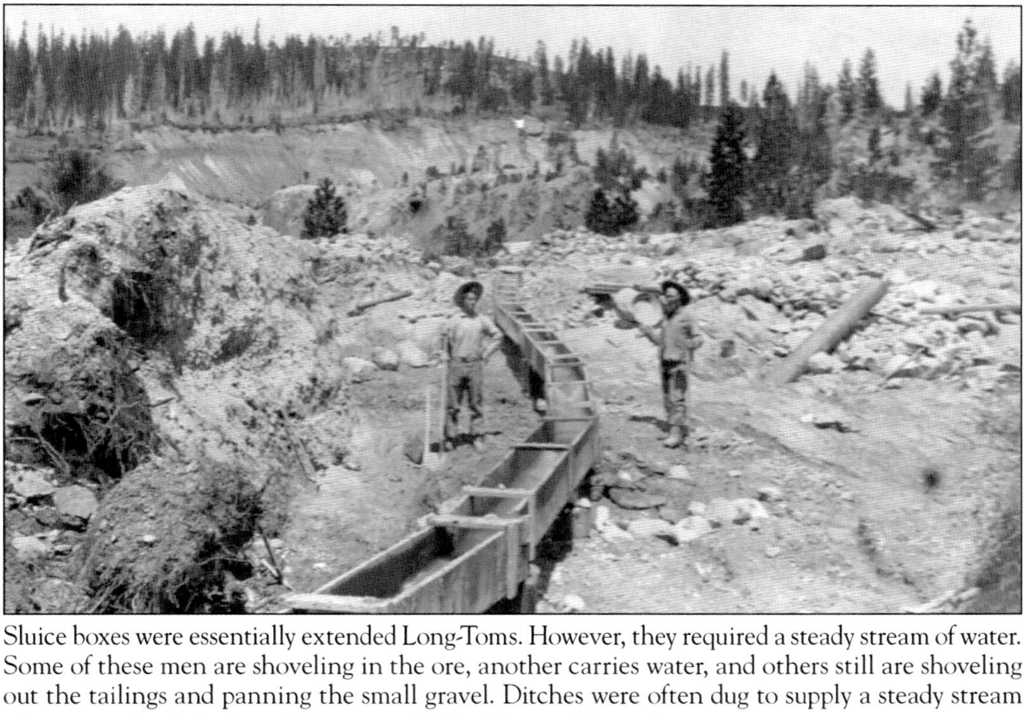

Sluice boxes were essentially extended Long-Toms. However, they required a steady stream of water. Some of these men are shoveling in the ore, another carries water, and others still are shoveling out the tailings and panning the small gravel. Ditches were often dug to supply a steady stream of water to the sluice box. (Courtesy of Yuba County Library.)

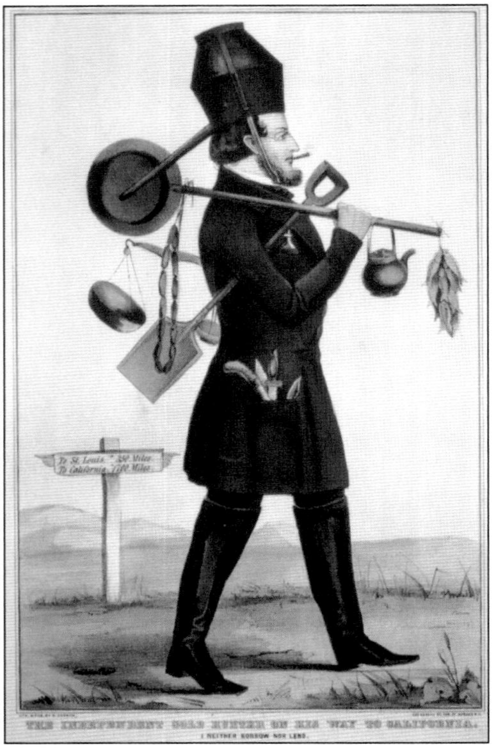

The Independent Gold Hunter on His Way to California is a drawing of a prospector carrying mining tools, cookware, and food as he walks to his claim. Not all miners came to California so well outfitted. Selling the necessary equipment, food, and clothing to the unprepared was as profitable as mining to those who had the foresight and the means to procure these goods. (Courtesy of Library of Congress.)

The Pioneers' Ten Commandments was written by James Hutchings and was first published in the *Placerville Herald* on June 4, 1853. It featured colorful, hand drawn scenes of mining life and humorous wisdom of the day, but it also offered good advice. It became so popular that it was later reprinted as a letter sheet and sold as stationery. (Courtesy of Library of Congress.)

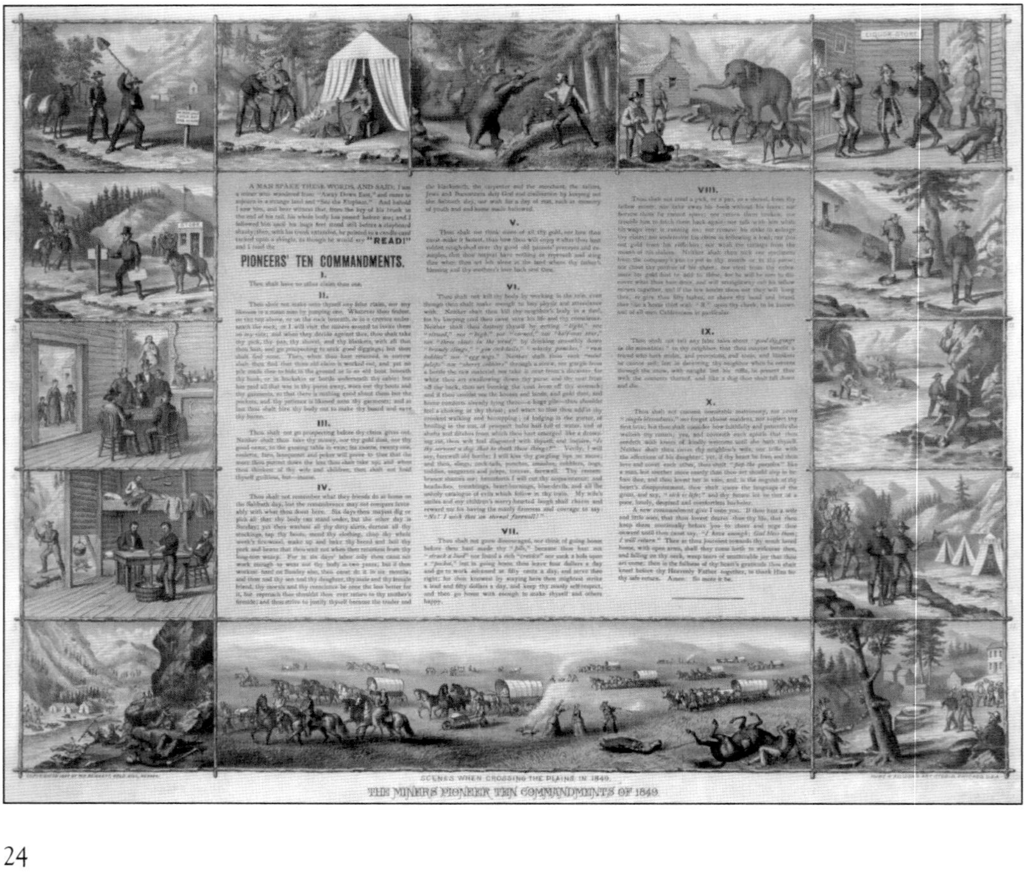

Two

Howdy Partner

Companies Formed to Improve Productivity

Early mining was exhausting and solitary work and it could be dangerous as well, often leading miners to forego comforts and necessities for the sake of making their pile. One could be injured or fall ill with no one to notice or care, or unscrupulous opportunists could take advantage of vulnerable loners. It was not uncommon to read stories in the newspapers about men found drowned on the river—their bodies unidentifiable because no one knew their names. For some, the life was too much and they returned to their families in the East.

Just as the miners had previously organized to establish rules, however, they eventually realized the benefits of working together and watching out for each other. Utilizing different methods to increase productivity meant they needed to form partnerships or companies and pool their resources. New companies began organizing before the members departed for the mines, making plans and bringing the appropriate skills and equipment with them. Often, group members came from the same ethnic background, religion, or place or origin. Many miners already in Yuba County organized into small companies, working together using more effective methods and employing new ideas to increase their productivity. With the evolution of the mining business in Yuba, the atmosphere was ripe for ambitious individuals who decided to stay. Some who caught gold fever were well-educated professionals like doctors or lawyers who were not necessarily cut out for the hard labor of mining but had skills that were in demand. Some hung out a shingle and practiced their profession, while others became involved in the process of establishing California politically—attending the constitutional conventions, seeking and holding office, or serving on the courts. Although many maintained an interest in a claim, they often diversified by opening a local service business, starting a stage line, or setting up shop by employing a previous skill like blacksmithing or carpentry. This helped the camps grow into towns capable of supporting the families that were arriving, and providing them with both necessities and comforts.

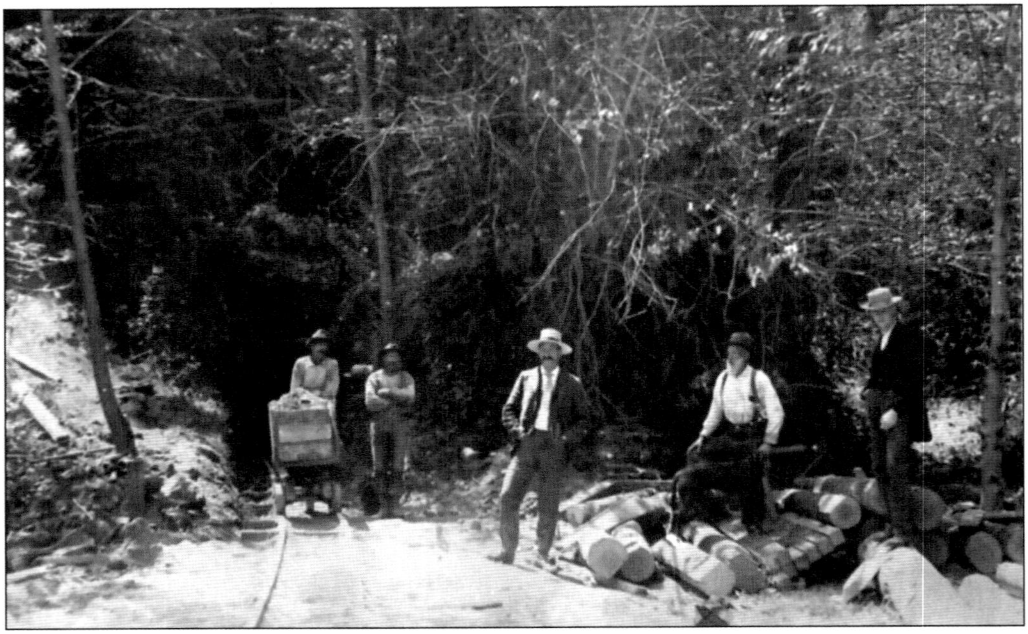

Gold was no longer being found in just the rivers and streams. Prospectors could dig a mine just about anywhere in the Yuba foothills—using dynamite to blast into the hills and then laying down tracks for ore cars. Shoveling lots of ore into the cars meant that there was a lot more to process. Prior to mechanization, men pushed carts themselves (like those above at the Miller Mine in New York Flat), or used mules (like Sam, below, in the Gold Bank Mine in Forbestown) to haul tools, men, and timber down and heavy loads of ore out. (Both, courtesy of Yuba Feather Museum.)

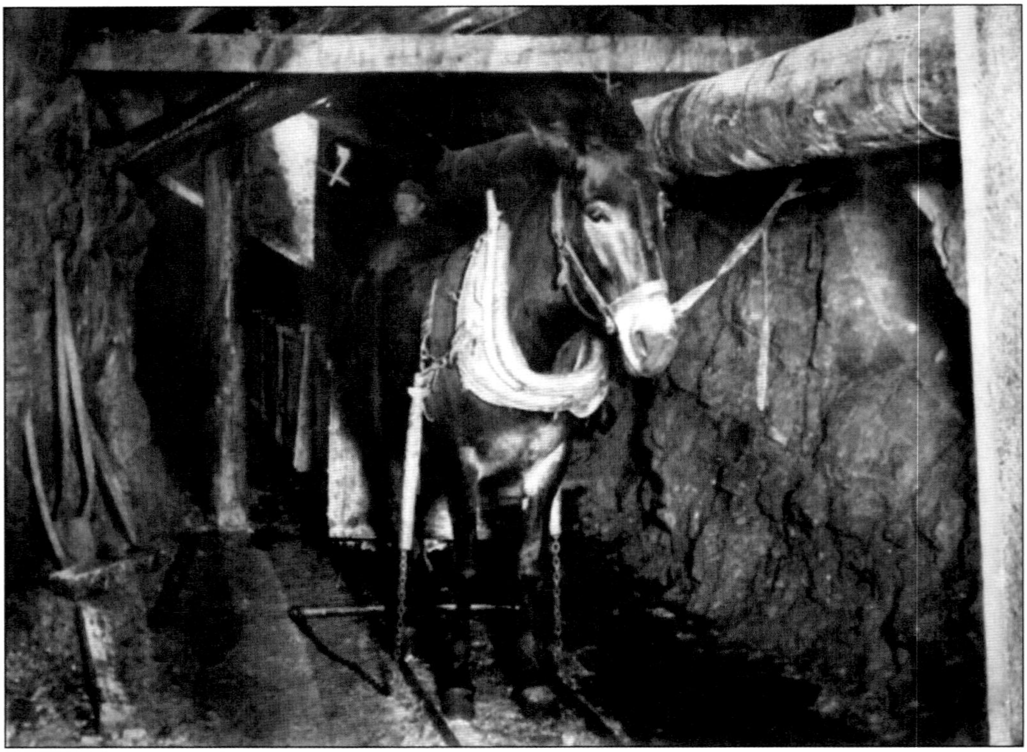

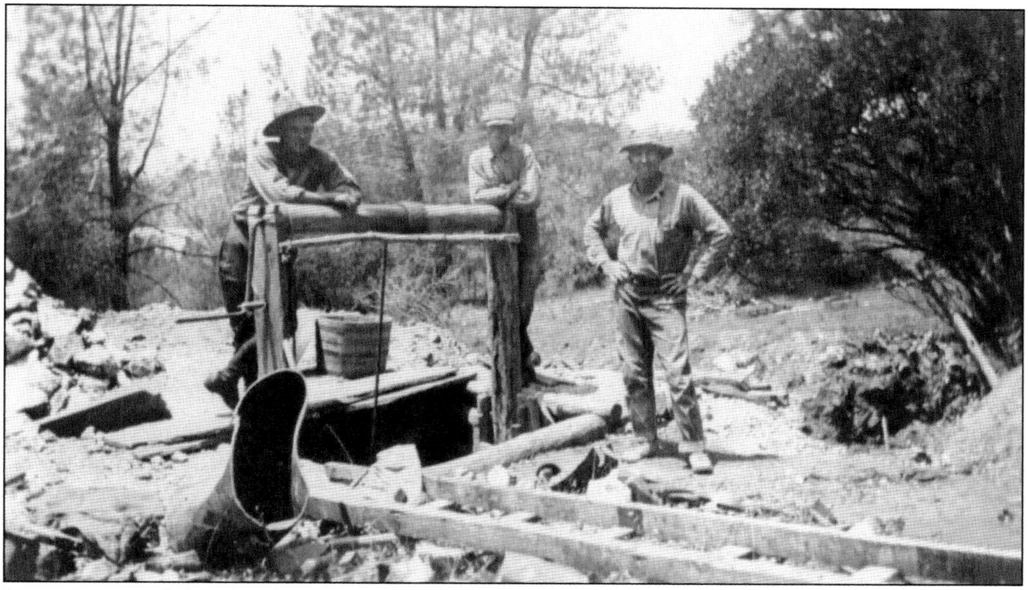

James Hapgood dug the shaft pictured above, and probably a tunnel. The ladder on the ground could be put in for men to climb down. Once inside, the miners would use a pick to break away chunks of ore and load it into the bucket. Someone on the outside had to crank up the bucket for further processing. This would have been considered a load mine, or a quartz mine, requiring miners to break the rock and separate out the gold. An arastra, like the one below, was likely close by for the next step of the process. The ore would be placed between millstones, and men or mules would turn the shaft to break the gold loose. It was then washed and amalgamated. (Above, courtesy of Marple/Hapgood Family; below, courtesy of Library of Congress.)

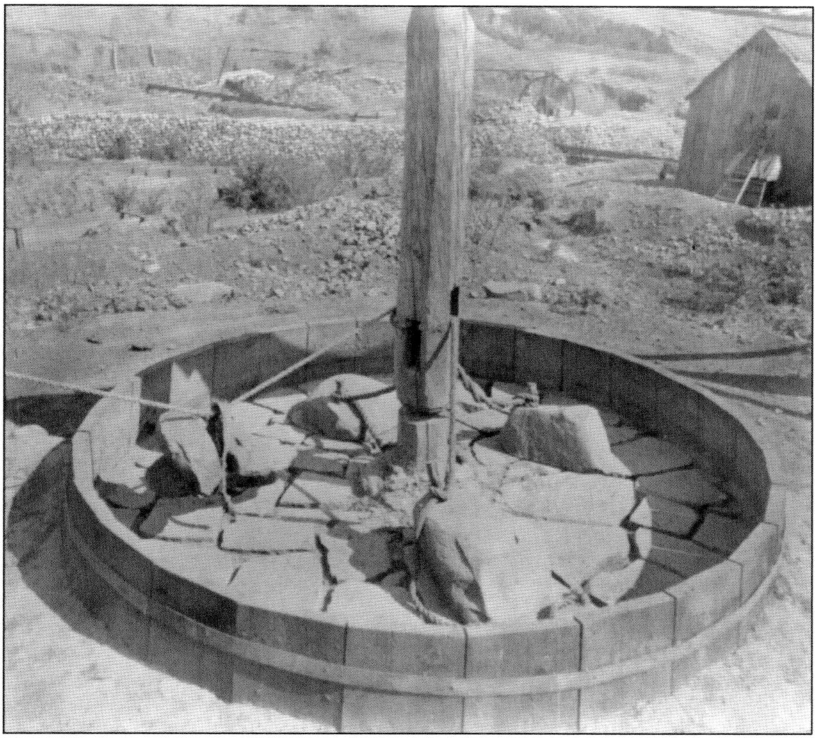

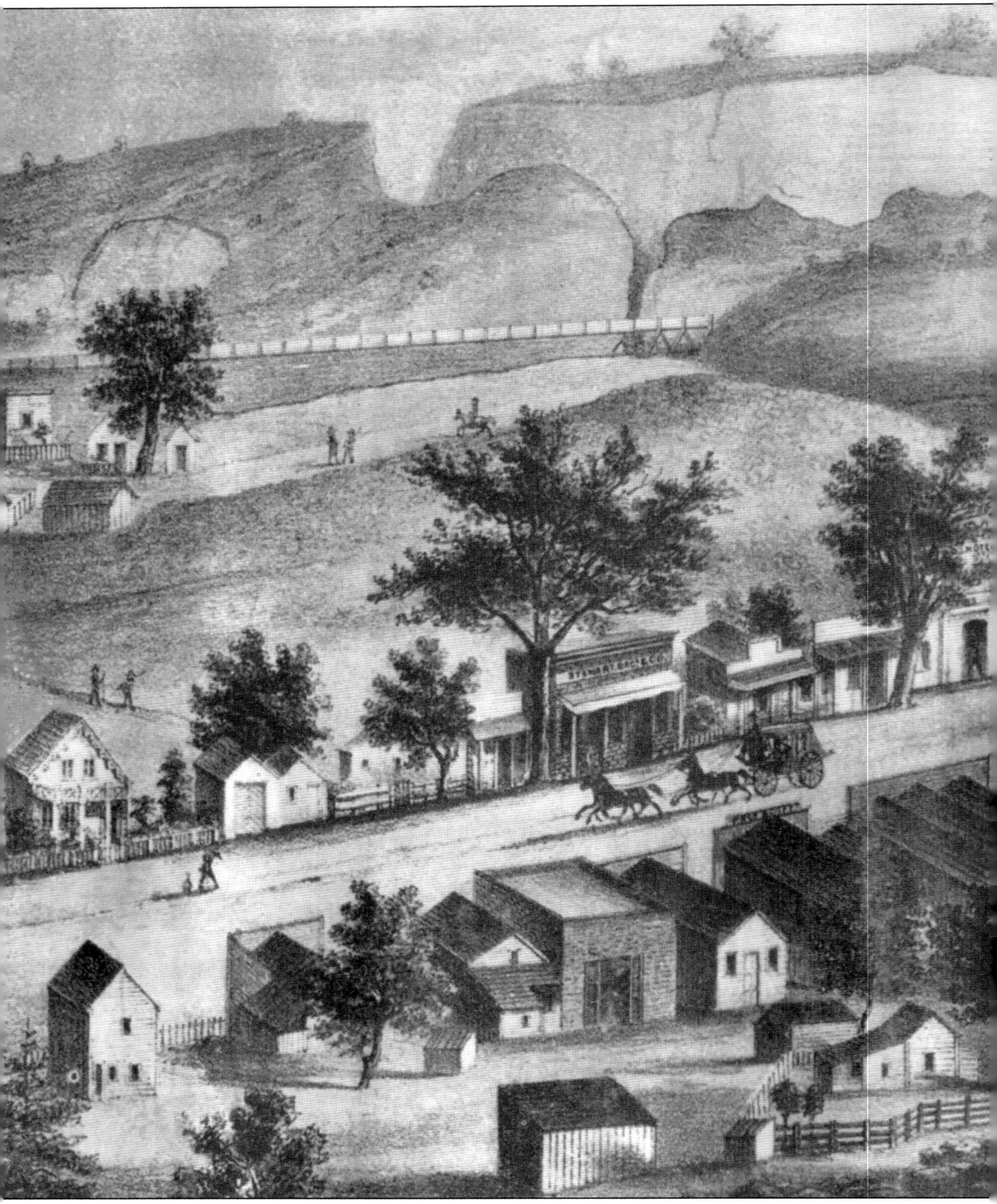

Richard Whitten Marple kept a store at Rose's Bar, and in his spare time, prospected in Big Ravine and along the creek. One day, he hit pay dirt, staked out a claim, and found it more rewarding than running the store. John Henry, the proprietor of another store and hotel at Rose's Bar, while reading a history of Africa one day, saw Marple come down the hill from his claim and said, "Here comes the Sultan of Timbuctoo." The name was first used to refer to the ravine, but the town

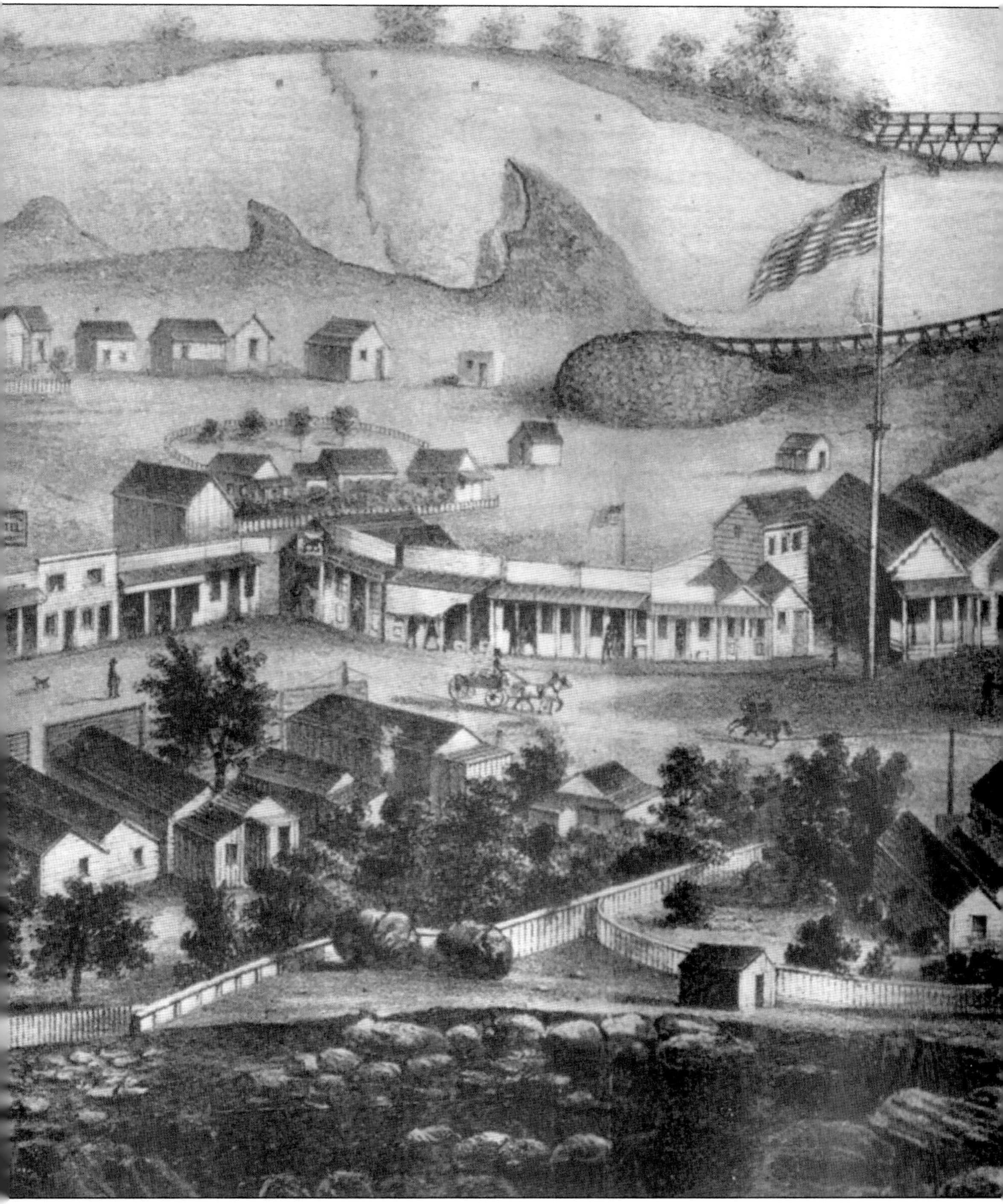

that grew there was named Timbuctoo as well. George Barrington's 1862 lithograph of Timbuctoo depicts the village that replaced Rose's Bar. The iconic Stewart Bros. & Co. store—which housed the Wells Fargo office—is featured prominently in the image as a stagecoach passes by it. The drawing is amazingly detailed and includes a number of flumes and sluices running through the background. (Courtesy of the Yuba County Library.)

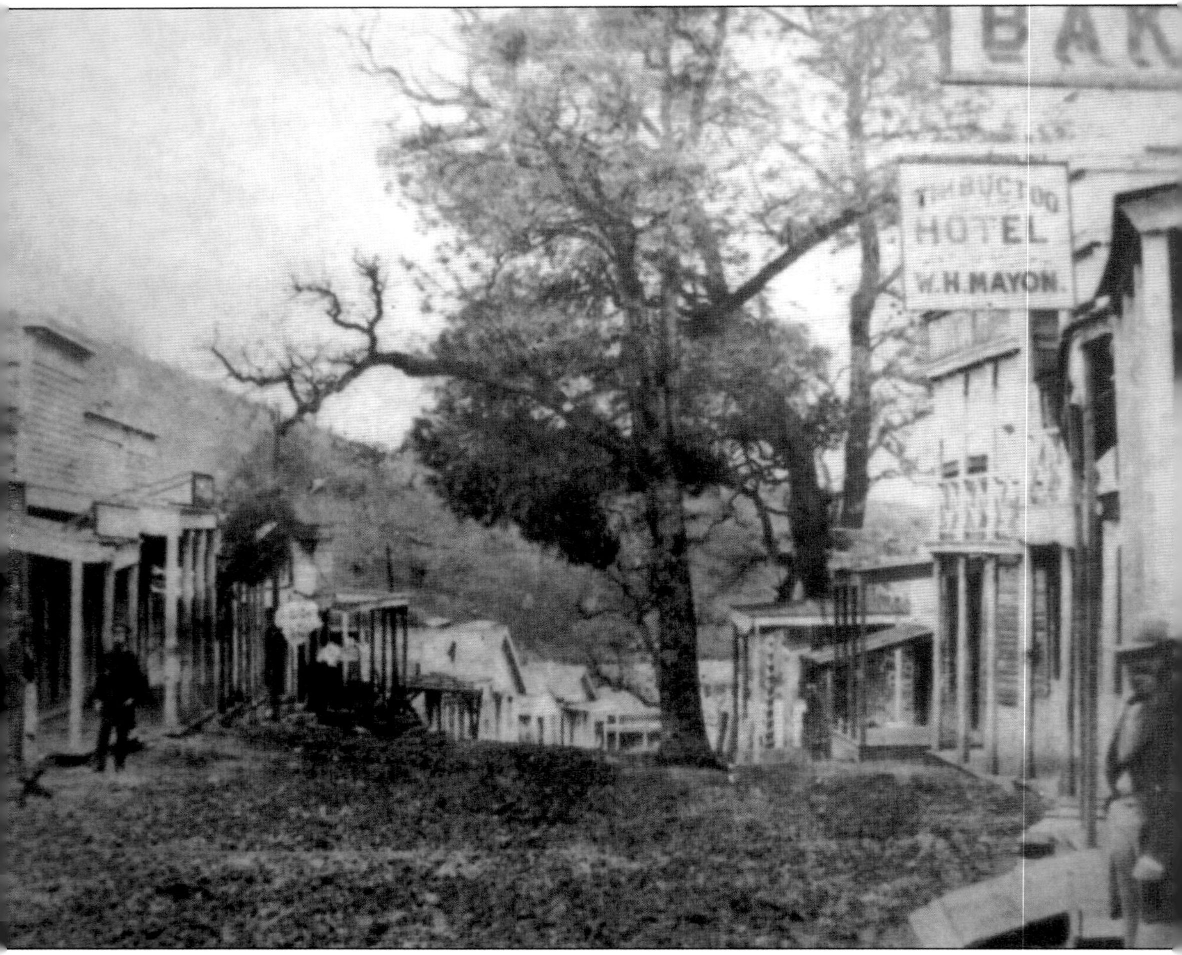

This photograph shows the town of Timbuctoo in 1859. The muddy, rutted road is steep here as it dips down towards the bridge that crosses the ravine. Hotel proprietor Thomas H. Mayon lived in town with his wife and four sons, while the bakery next door belonged to James and Caroline Pine, who had a five-year-old son, William. Timbuctoo hosted many homes, stores, hotels, and saloons, as well as a theater and telegraph office. (Courtesy of Library of Congress.)

Richard Whitten Marple was one of the first to secure a claim in the Timbuctoo Ravine, as shown on this map drafted from survey data taken in May 1859. Marple married and had a large family, which remained in Timbuctoo for several generations. Below, two of Marples's sons are seen using a sluice box on the family claim. (Both, courtesy of the Marple/Hapgood Family.)

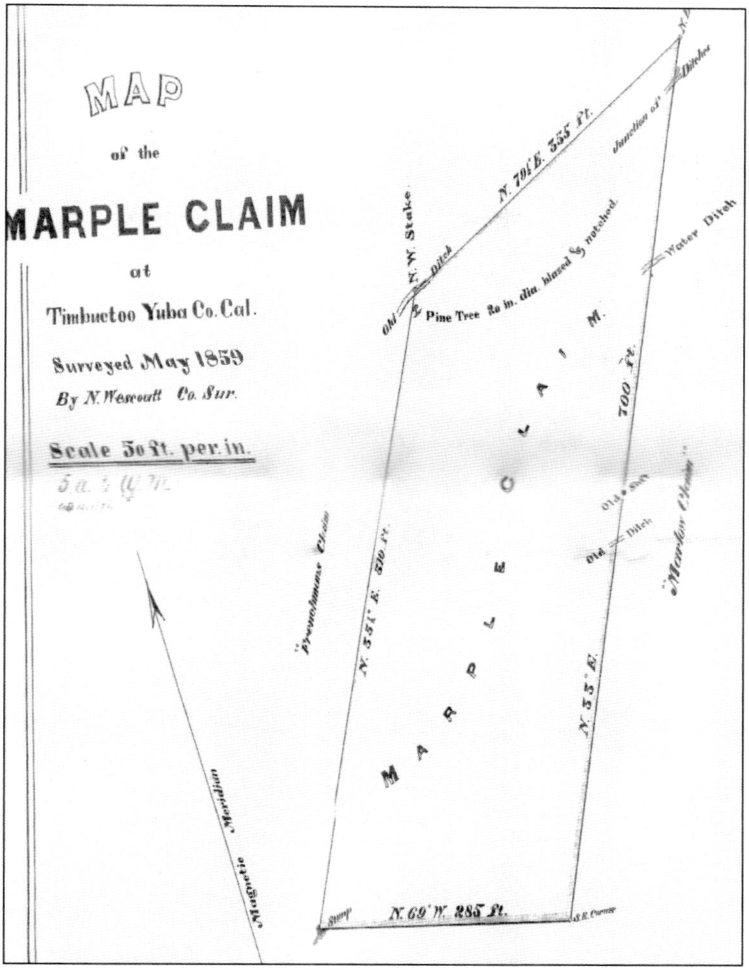

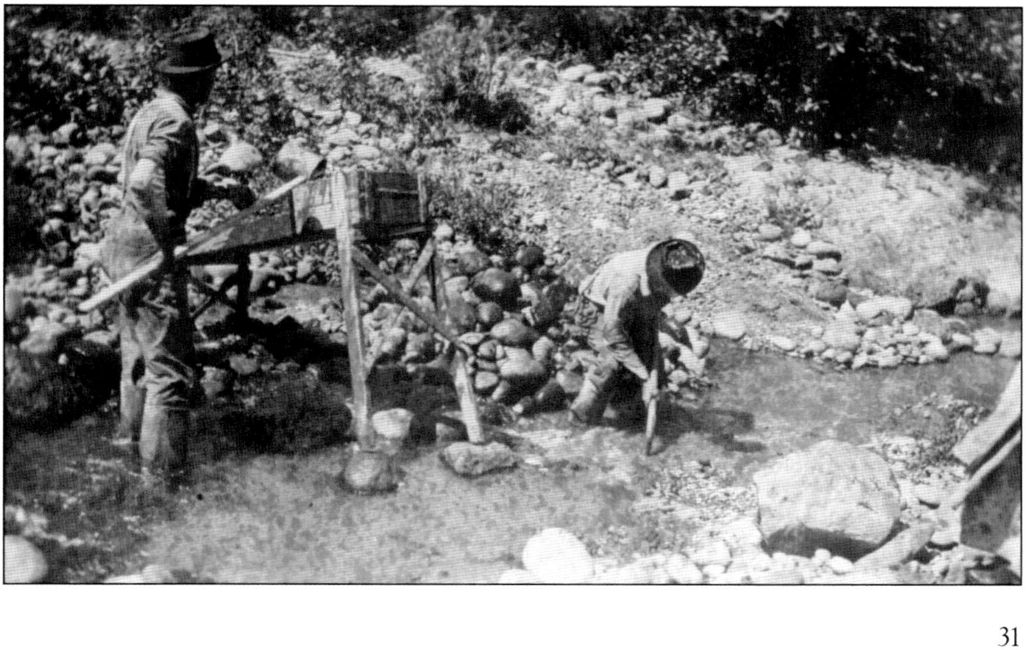

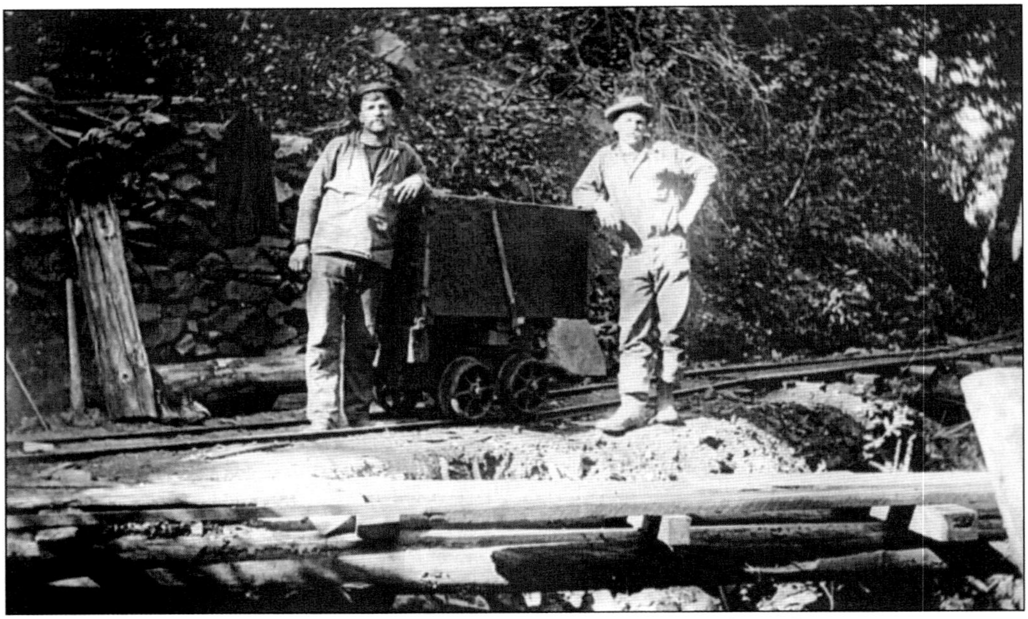

Henry and George Beik were two of the 12 brothers who owned the Beik Brothers Mine. George was killed in a cave-in on February 28, 1900, and his mother forbade any of her other sons from returning to the mine. Later however, the mine was reopened by the Lewis family. The mine was located on Mosquito Creek, on Mosquito Ridge, near Forbestown. (Courtesy of Yuba Feather Museum.)

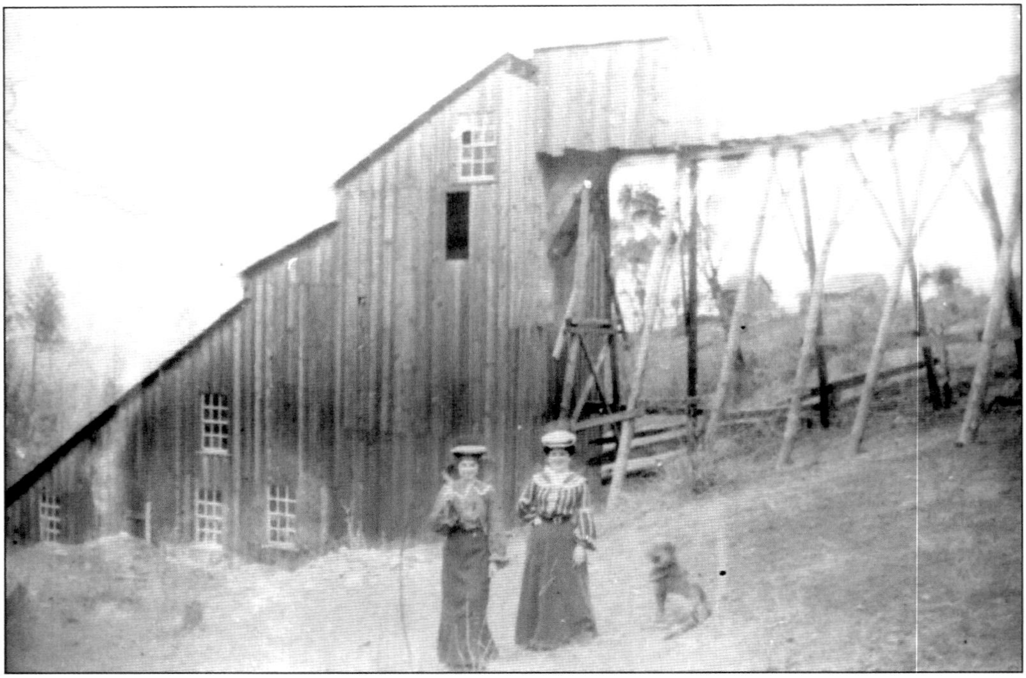

Two unidentified women pose in front of the Santa Rosa Mine in Rackerby, which housed a water powered mill for crushing gravel. Improved methods for handling gravel rapidly made mining more lucrative. A drift mine is one worked with a horizontal opening to access gravel deposits containing gold, as opposed to solid rock like quartz or granite containing gold. (Courtesy of Yuba Feather Museum.)

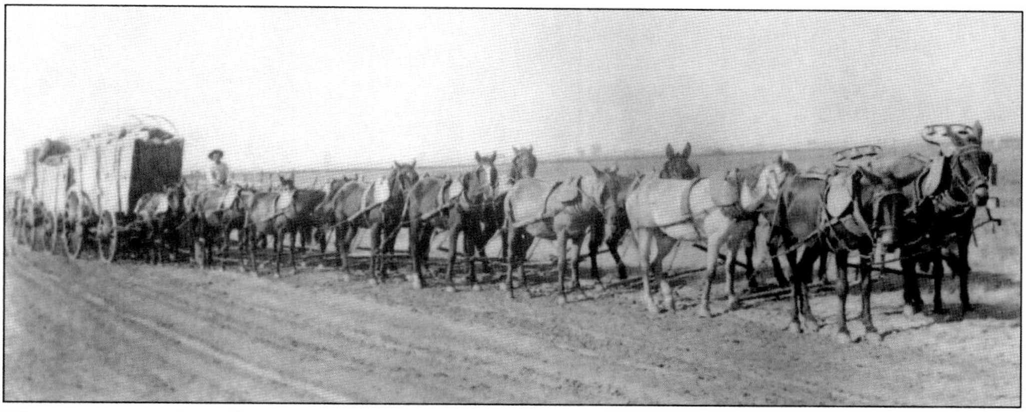

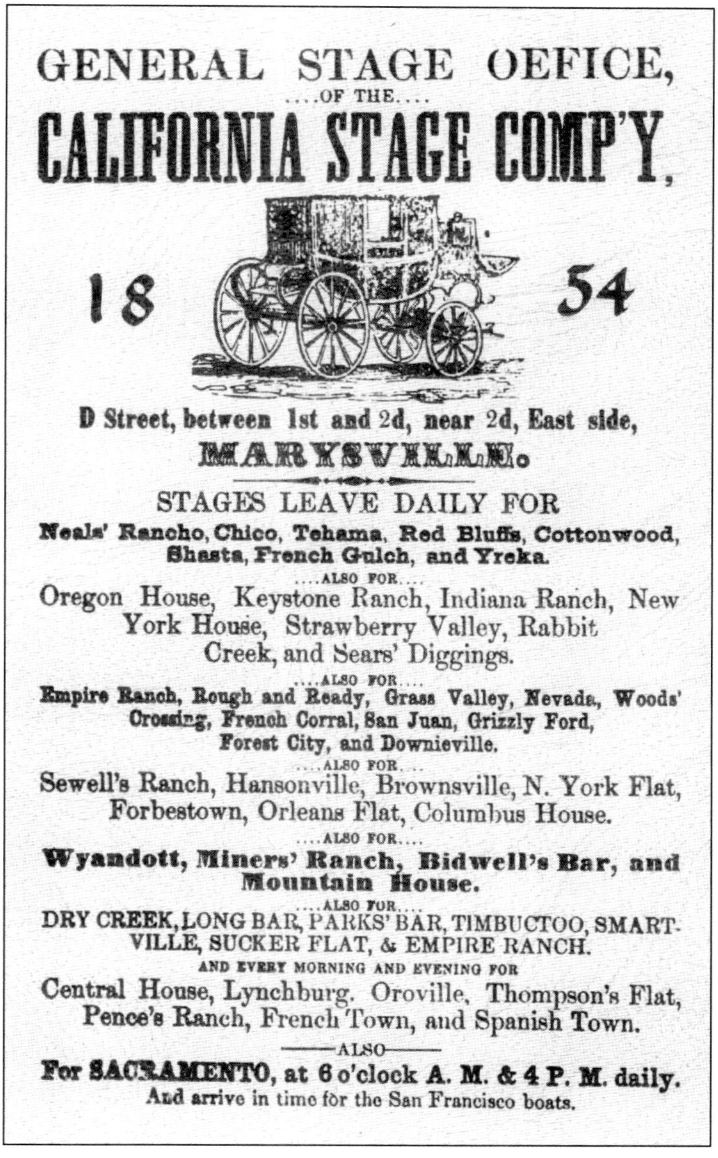

The transportation of supplies and people to the mines fueled the development of roads in the area, as well as stage and freight companies. At right, a stage schedule from 1854 lists many of the mining camps that the coach stopped at. The stage also carried mail and newspapers to these places and was an important means of communication. Freight was delivered on wagons pulled by long-line teams, like the one owned by Andrew Kneebone pictured above around 1880. (Above, courtesy of the Penn Valley Chamber of Commerce; right, courtesy of Yuba County Library.)

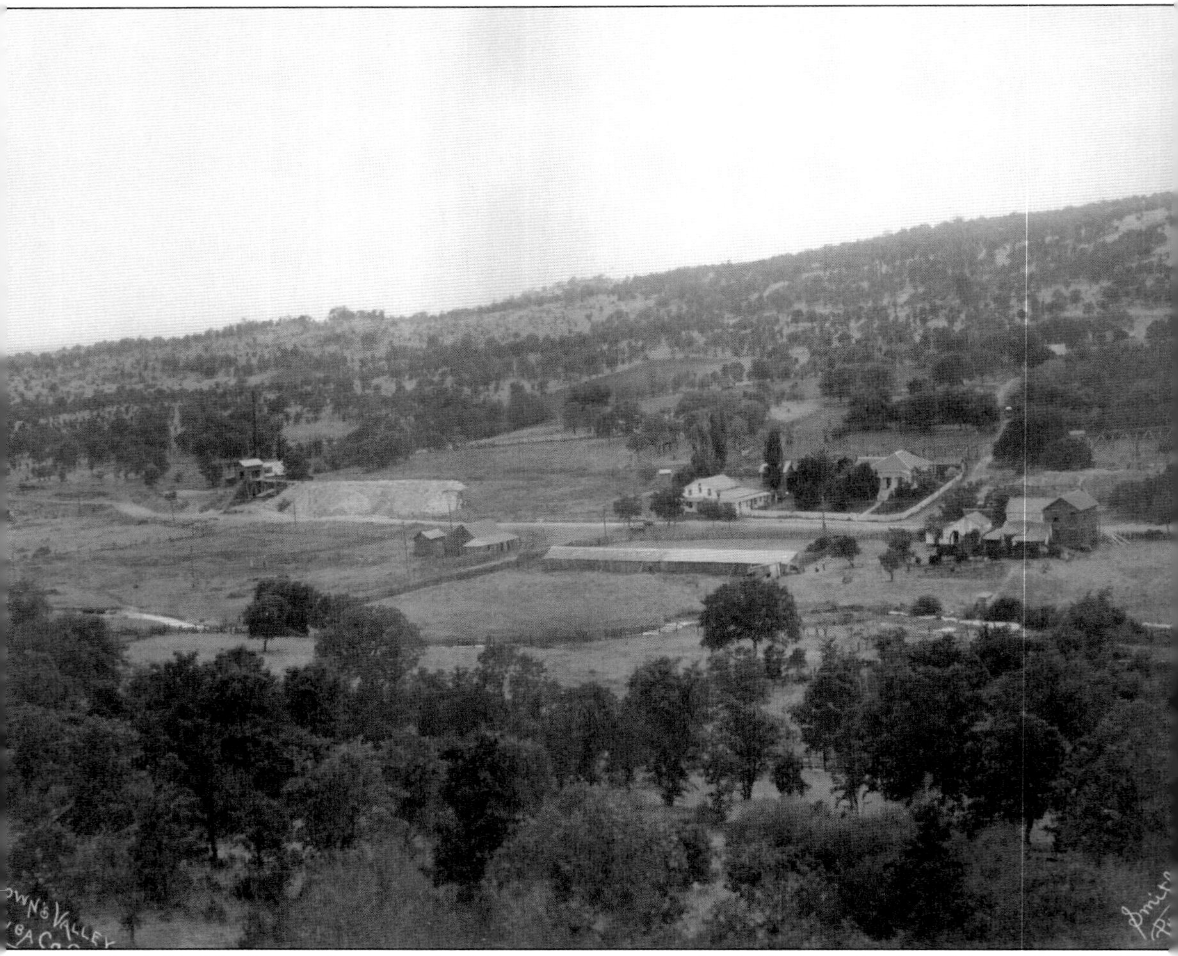

In 1850, Mr. Brown discovered gold-bearing quartz in the fields 12 miles from Marysville. This place became a lucrative hard rock mining area and is now called Brown's Valley. Typical of other Yuba mining settlements, people lived in a town that was directly above the mines. In 1852, a water-powered wheel was erected for a mill on the west side of a dry creek. The water source did not provide enough water to power the mills, and there were no other substantial water sources nearby, so water was shipped to the area at great expense. (Courtesy of Yuba County Library.)

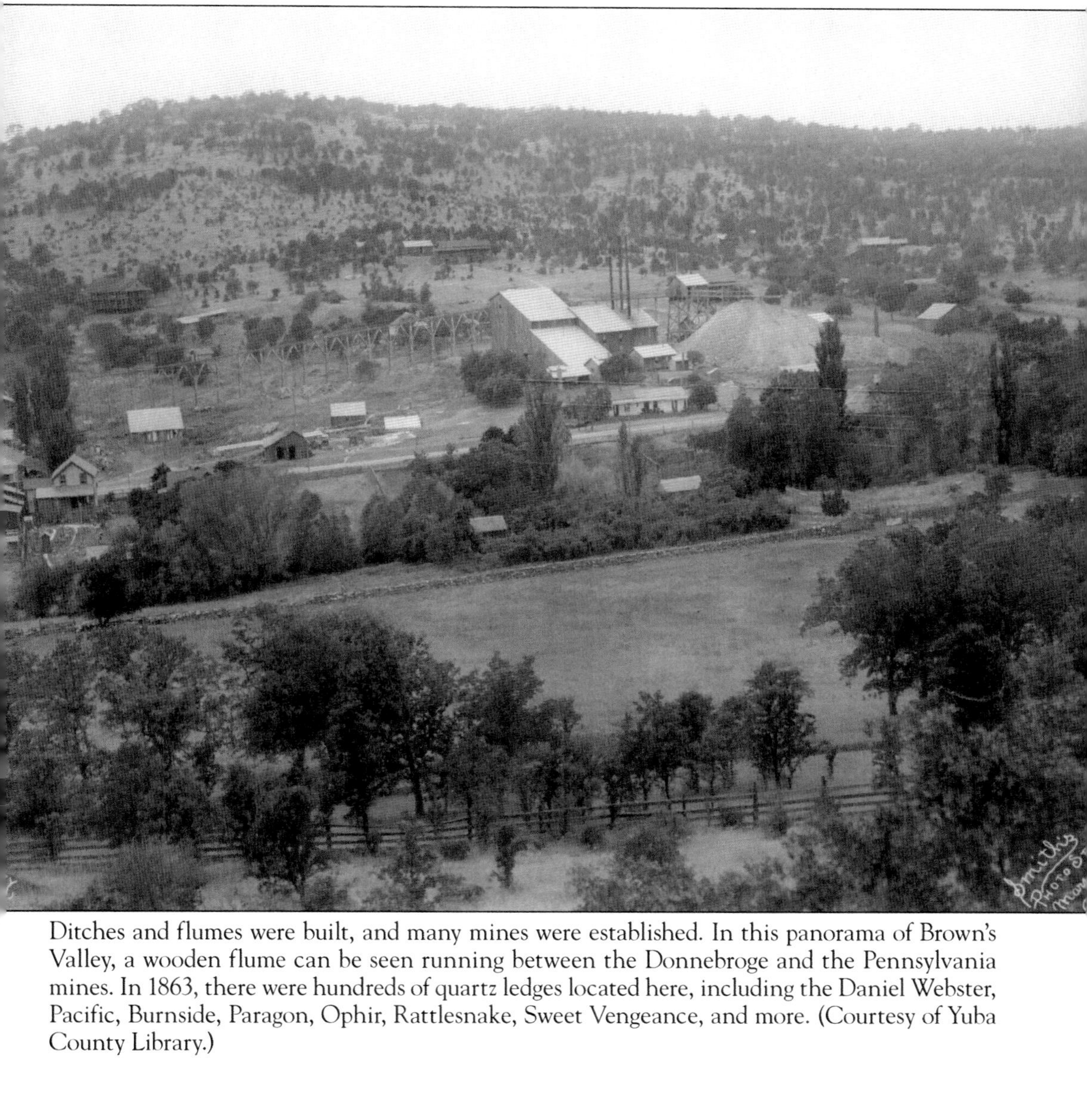

Ditches and flumes were built, and many mines were established. In this panorama of Brown's Valley, a wooden flume can be seen running between the Donnebroge and the Pennsylvania mines. In 1863, there were hundreds of quartz ledges located here, including the Daniel Webster, Pacific, Burnside, Paragon, Ophir, Rattlesnake, Sweet Vengeance, and more. (Courtesy of Yuba County Library.)

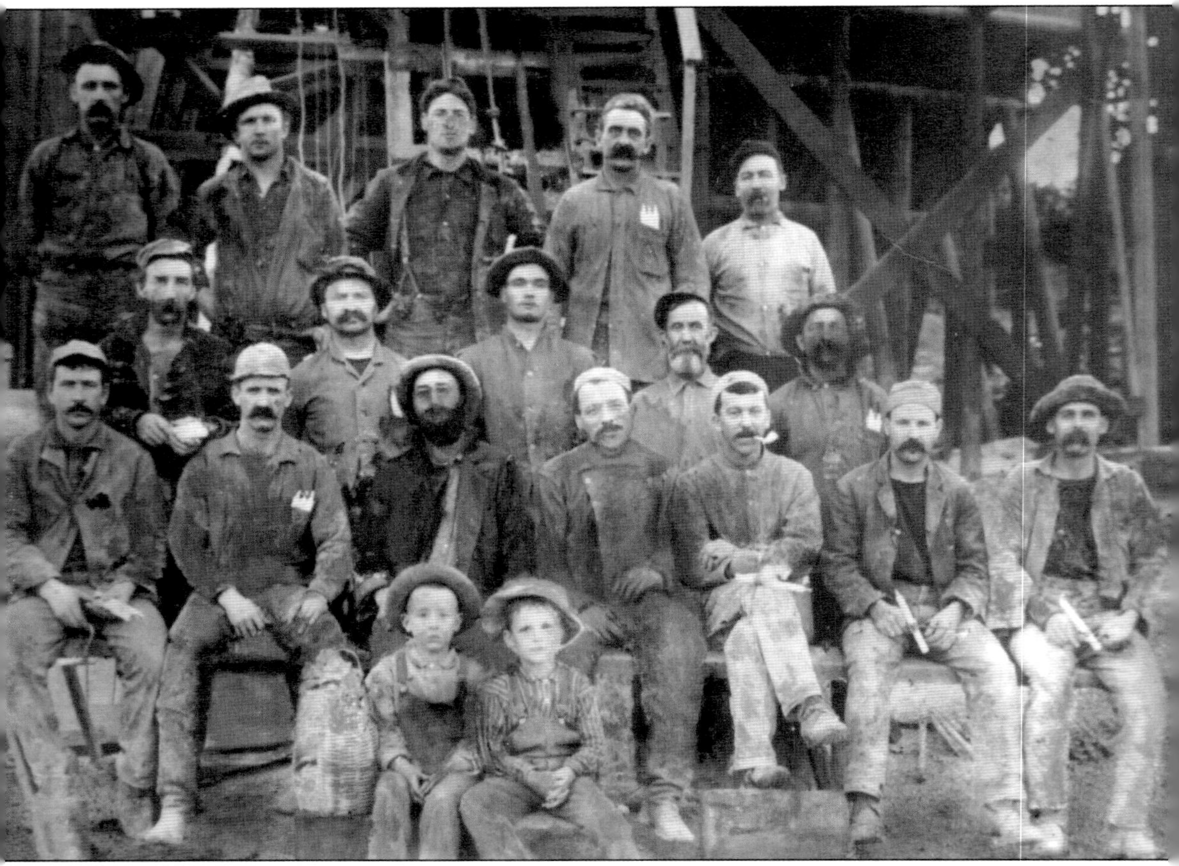

Miners from the Pennsylvania Mine pose in 1902 before going underground. They are holding the candles that will light their work. Jacob Sperbeck, age 26, is seated at the far left of the third row. The children in the first row were likely helpers and were not yet protected by child labor laws. (Courtesy Roberta Sperbeck D'Arcy.)

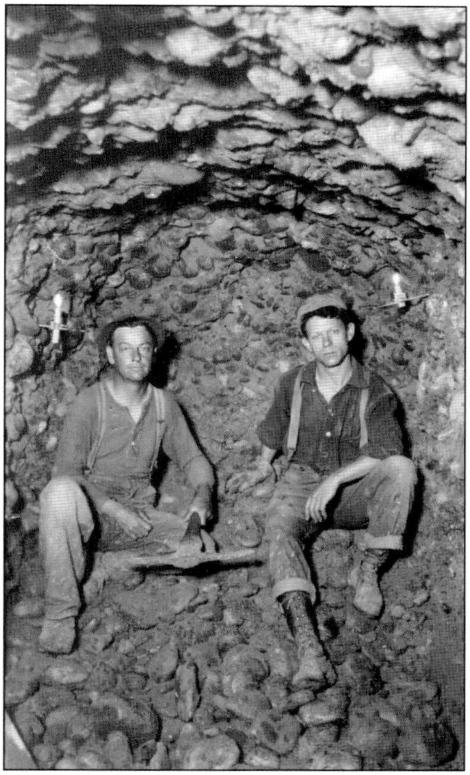

These photographs provide two views of men working underground in the Browns Valley mines. Working in these small spaces with only a lamp for light must not have been very pleasant. Below, Mr. Reed and Mr. King pose in the Pennsylvania Mine. (Both, courtesy of Ivadene Leech.)

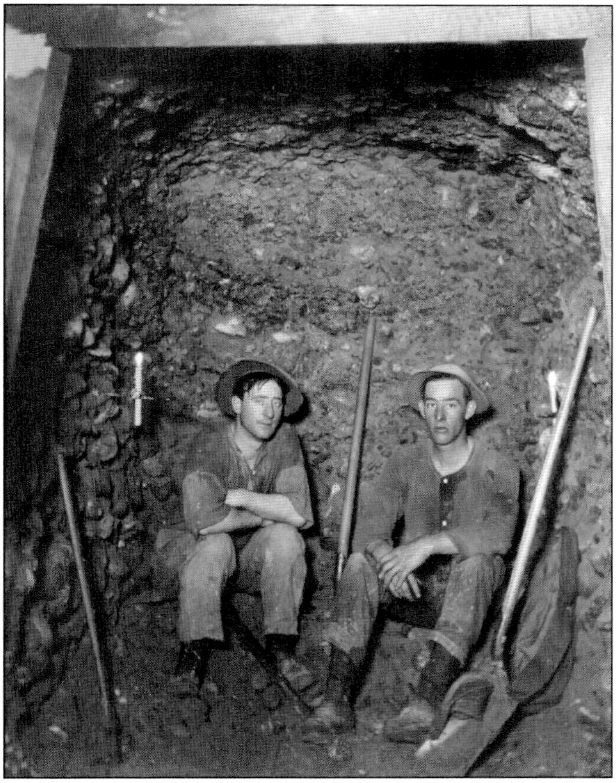

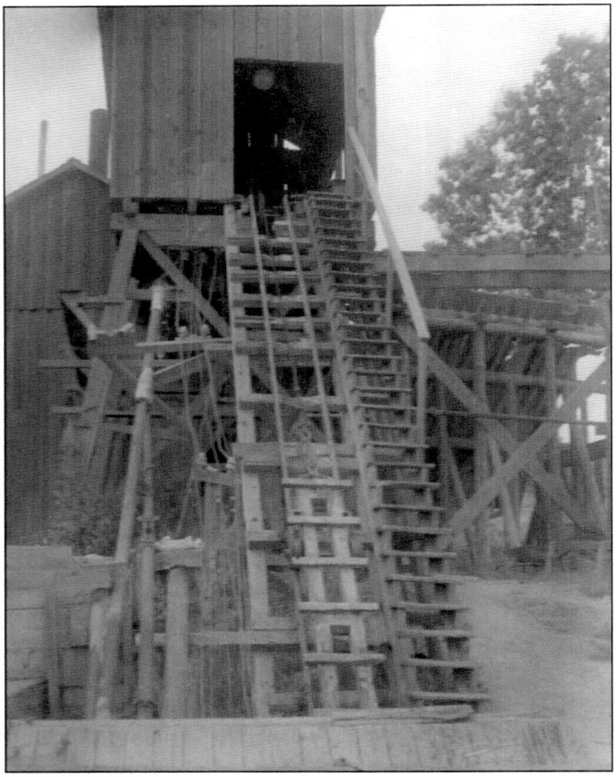

By 1867, the Pennsylvania Mine reached 600 feet down into the earth, and had run drifts of 200 feet. It also had a 16-stamp steam-powered mill. The photograph at left shows the hoist house of the Pennsylvania in 1867, while below, ore cars are seen coming out of the mine. (Both, courtesy of Ivadene Leech.)

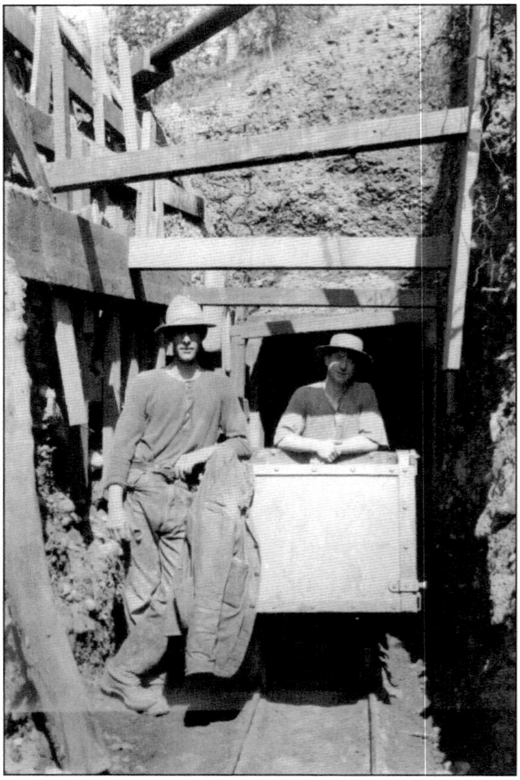

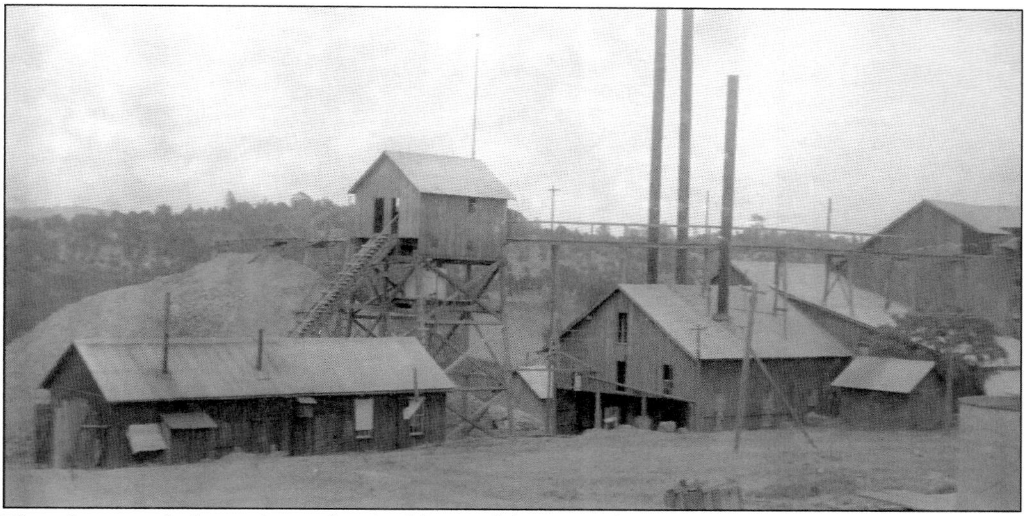

This photograph shows the surface buildings of the Donnebroge Mine, which in 1867 had already dug down 500 feet and drifted 200 feet. It operated an eight-stamp steam-powered mill. (Courtesy of Ivadene Leech.)

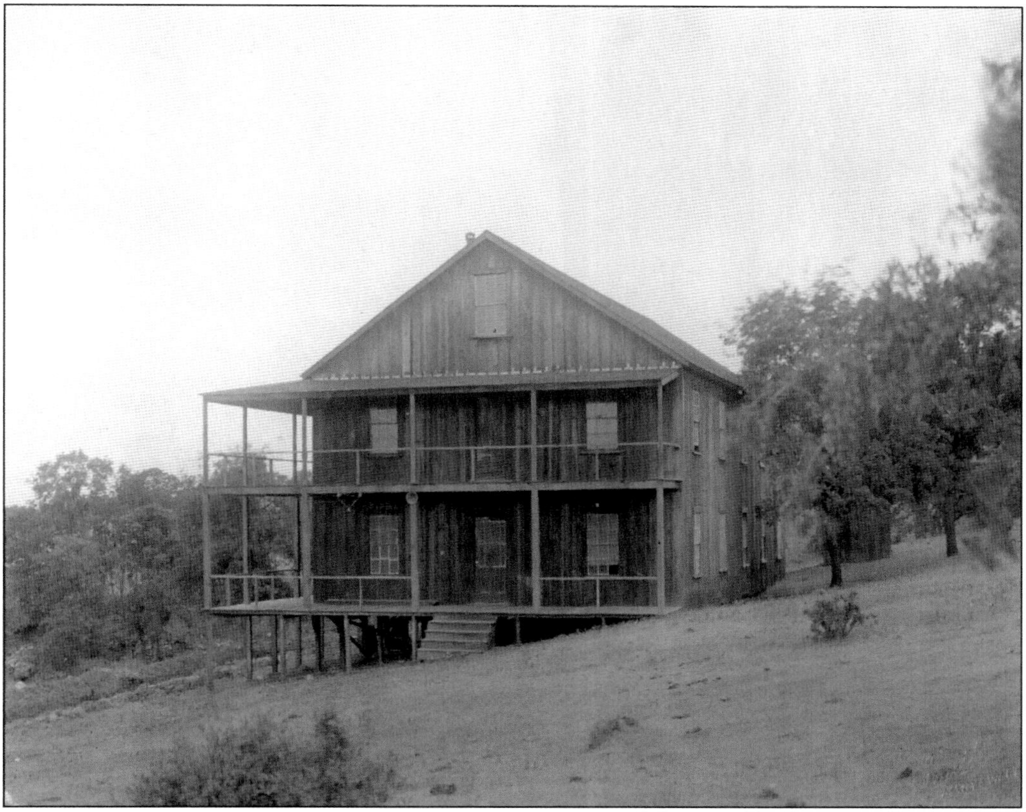

Many of the mines had bunkhouses for single miners. This one was at the Donnebroge Mine. Such buildings were one way to attract and keep laborers for the mines. (Courtesy of Ivadene Leech.)

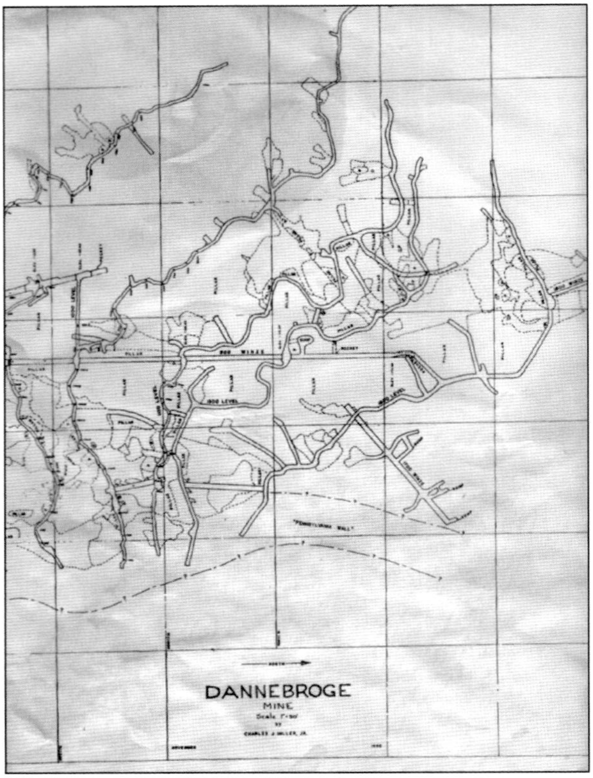

This diagram of the Donnebroge Mine, by Charles J. Miller Jr., shows the extent and depth of the tunnels. Notice how close it is to the Pennsylvania Mine. One local resident said that the whole town was undermined by the deep tunnels. (Courtesy Charles L. Miller Jr.)

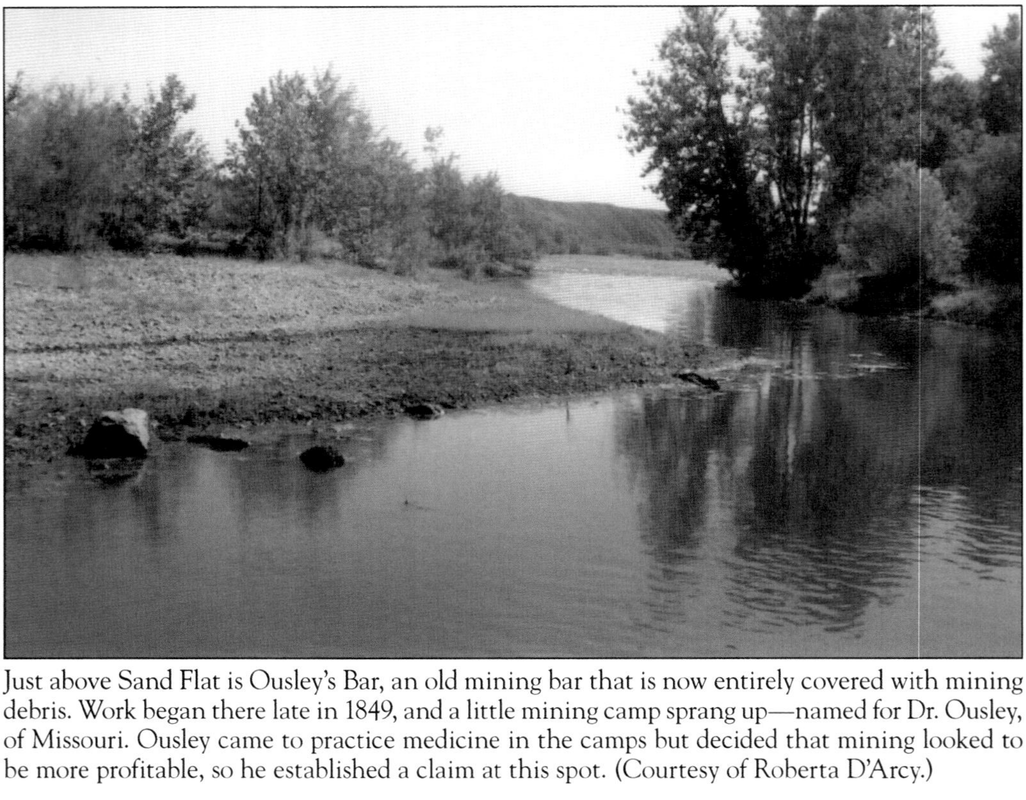

Just above Sand Flat is Ousley's Bar, an old mining bar that is now entirely covered with mining debris. Work began there late in 1849, and a little mining camp sprang up—named for Dr. Ousley, of Missouri. Ousley came to practice medicine in the camps but decided that mining looked to be more profitable, so he established a claim at this spot. (Courtesy of Roberta D'Arcy.)

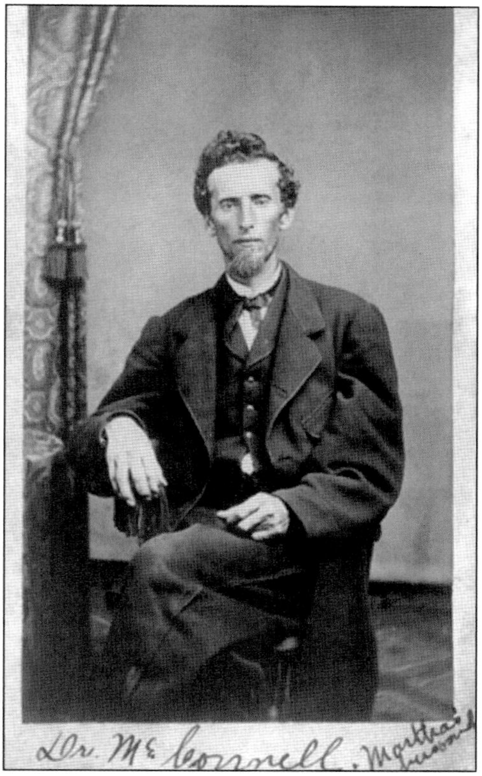

Dr. James McConnell, born in Pennsylvania in 1825, came overland to California with his brother and mined in Downieville. His wife, Martha, and infant son, John Taylor, soon joined him there. They had two more children and moved to Smartsville, where James was the town physician and Martha was the druggist. When he passed away, Martha remained to raise her family—supported by her profession. (Both, courtesy of Linda Lutton Jackson.)

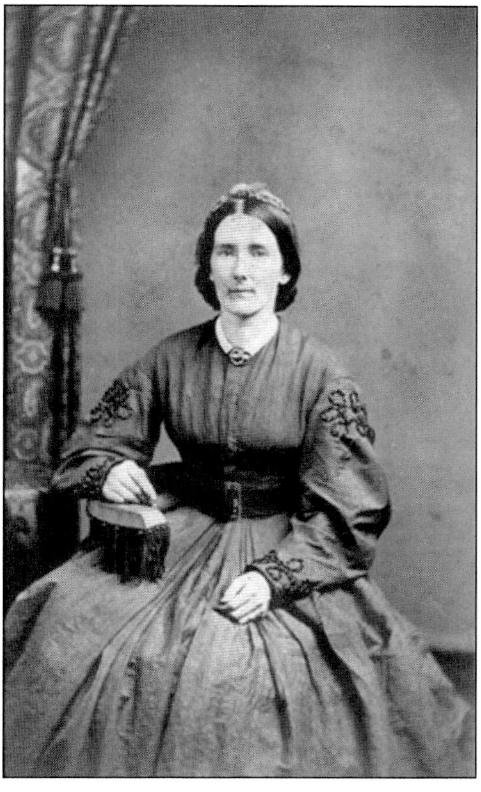

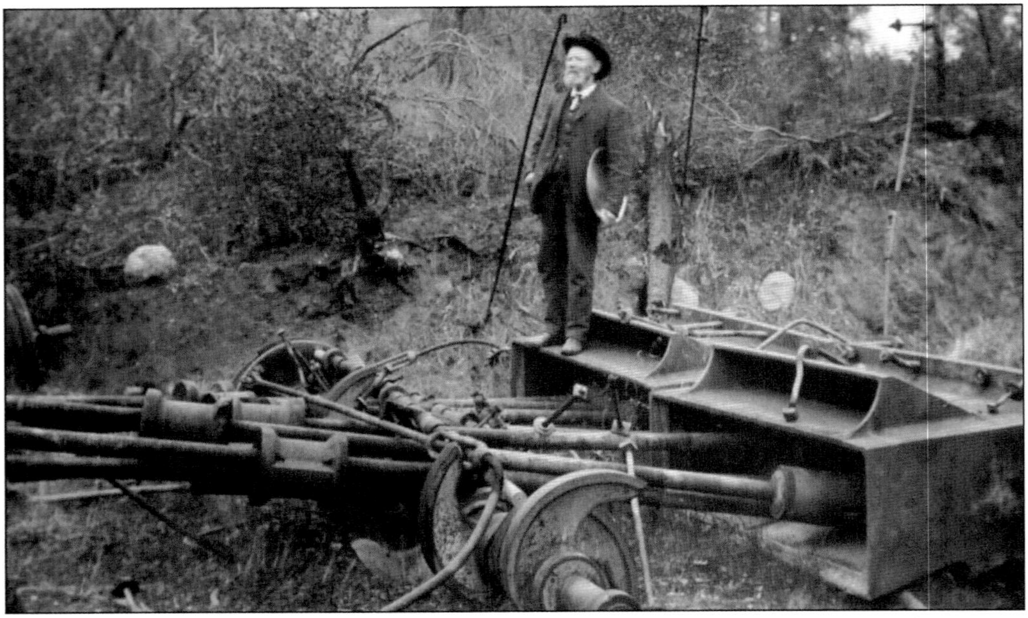

Samuel Owen Gunning was born in Sligo, Ireland, in January 1834. When he was only 12 years old, he left home and found employment on a coasting vessel for several years. He arrived in San Francisco in October 1857 and came directly to Smartsville. For a time, he was superintendent of the Excelsior Water and Mining Company, and for 10 years, was in charge of the Timbuctoo property. He held office in Yuba County as both auditor and recorder for over 30 years. He is pictured here surveying an area while standing on one of his stamp mills, which has been vandalized. (Courtesy of the McCay Family.)

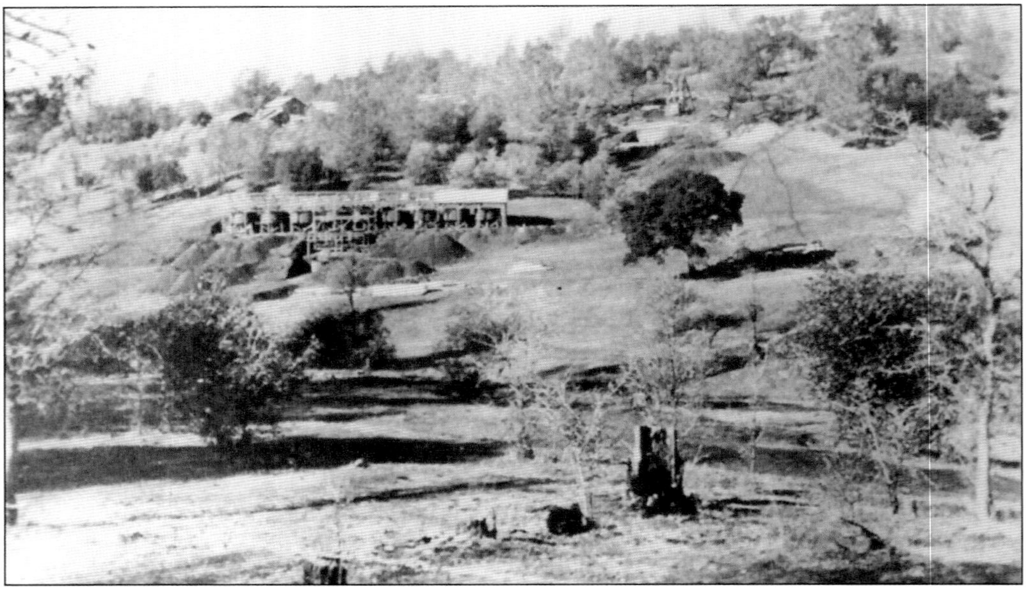

The Marc Anthony claim was located in Timbuctoo, high upon a peak above the Yuba River. In 1863, a 10-stamp mill was erected, and a shaft was sunk to 300 feet. Ore cars came out from the side of the hill and ran directly to the stamp mills, which crushed the ore and performed the initial extraction process. The remaining ore was washed to a cyanide plant, which further processed it. (Courtesy of Mary Clark.)

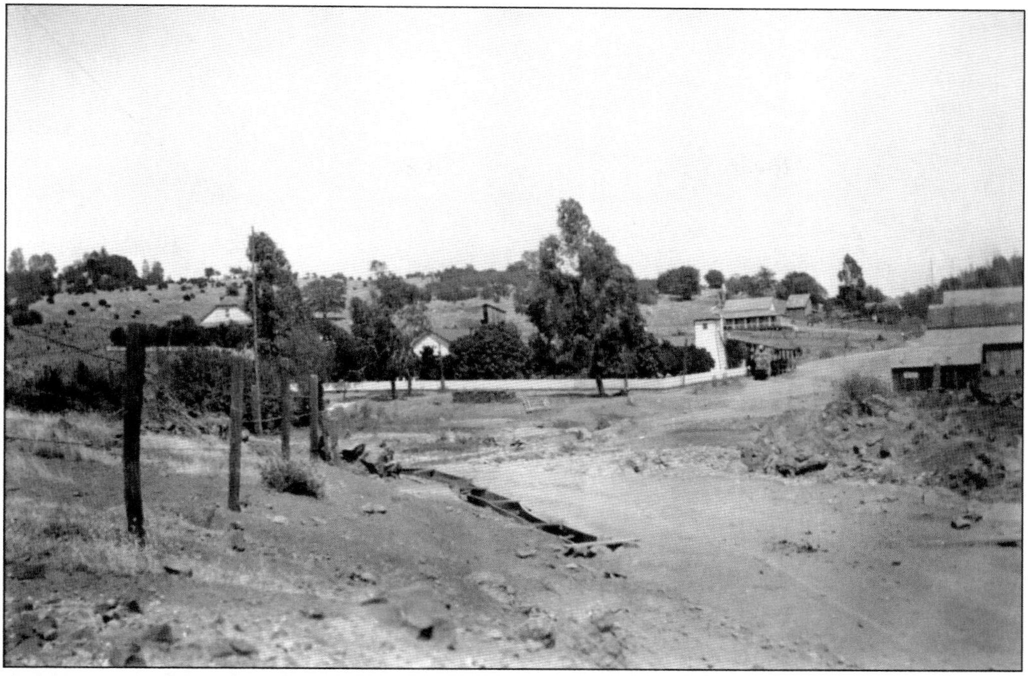

Gold was not the only valuable mineral found in Yuba. In 1863, copper was discovered in the vicinity of Timbuctoo. The rush of copper prospecting that followed led to the discovery of valuable quartz lodes. The most prolific copper mine in the area was the Lone Tree Ledge. Originally known as the Well Lode Copper Mine, the lode was first discovered during the digging of a well on Purtyman's Ranch in what is now Spenceville. It was locally operated for a short time, but the San Francisco Copper Mine and Reduction Works Company took over and sank shafts and drifts to a depth of roughly 150 feet. After a cave-in in 1880, the company excavated an open cut mine 300 feet long, 70 feet wide, and 75 feet deep. The mine produced more than 150,000 tons of ore. (Both, courtesy of the Nevada County Historical Society.)

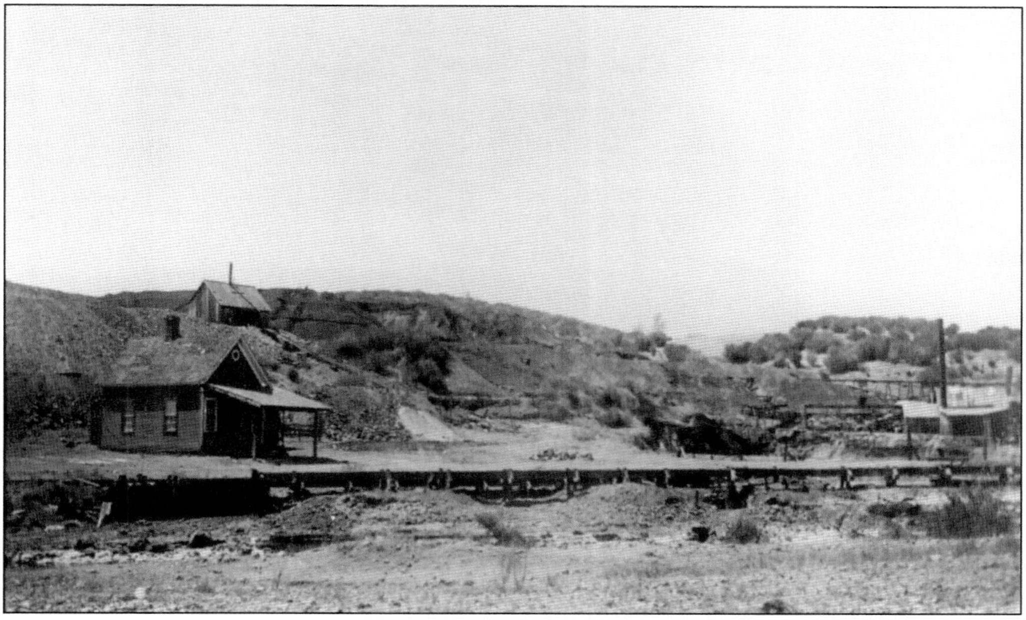

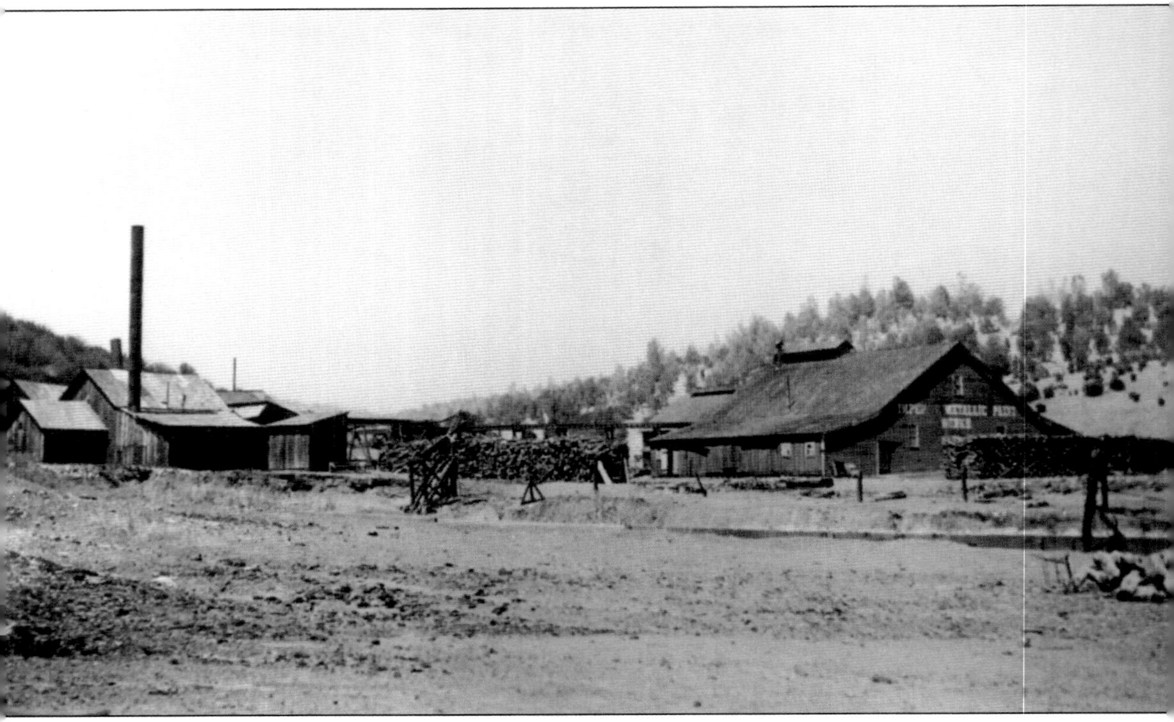

The Imperial Paint Company reworked the tailings left behind by the San Francisco Copper Mine and Reduction Works Company, recovering additional cement copper through leaching. It also used the iron oxides produced—a residue of the roasting and leaching process—as a Venetian red paint pigment. The use of the pigment was discontinued when it was discovered that the paint caused nails to rust. (Courtesy of the Nevada County Historical Society.)

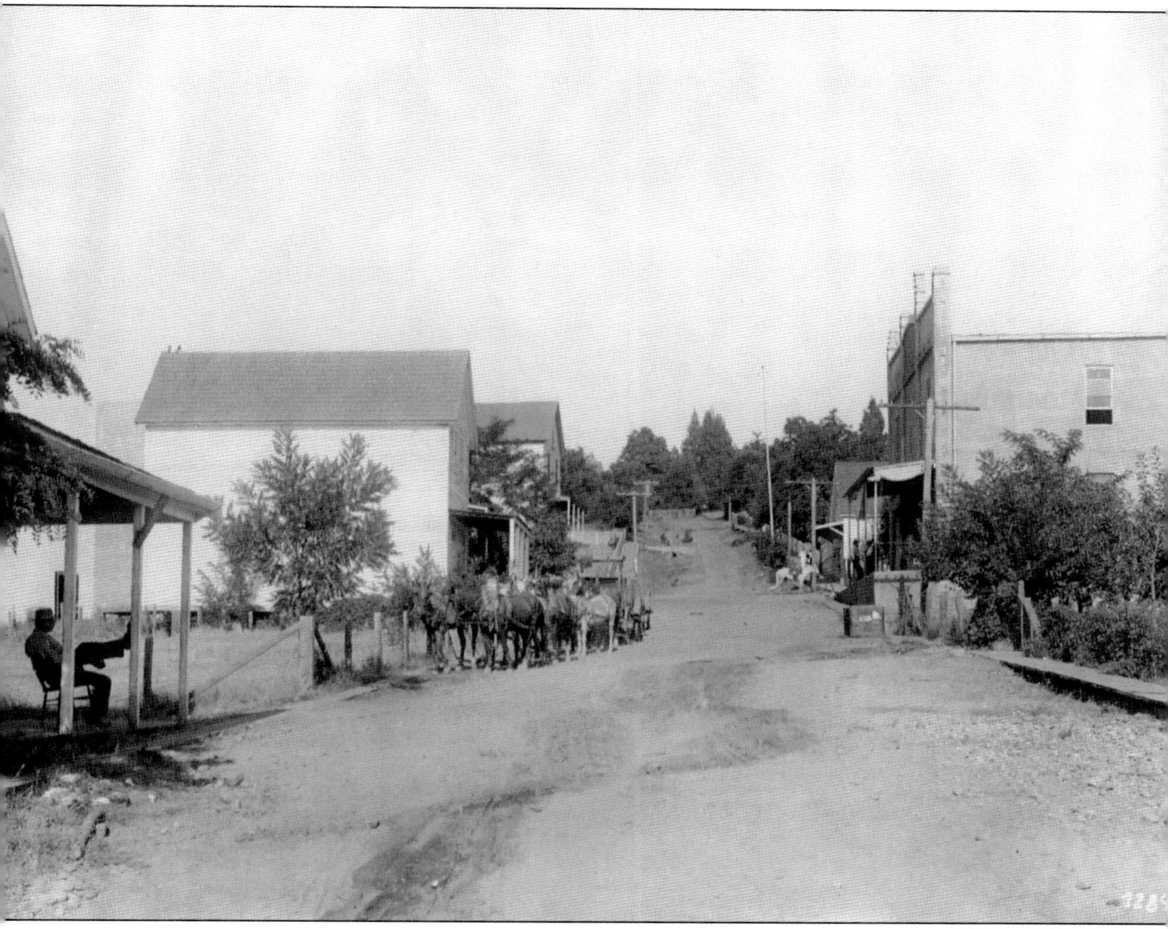

Located on a ridge between the North Fork and Middle Fork of the Yuba River, Camptonville was founded by miners in 1852 and was a significant mining town and one of the most northern towns in Yuba County. Throughout the area's many boundaries changes, however, it has remained a part of Yuba County. In its time, it was a hub for the supply chain that supported the far northern mining towns. It hosted many services and entertainment venues such as saloons and even a bowling alley. (Courtesy of Yuba Feather Museum.)

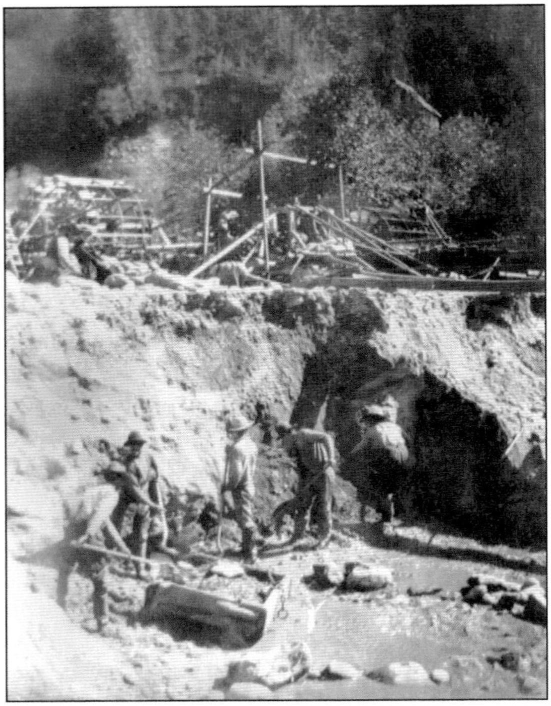

Elevating water wheels date to the 16th century. These wheels were constructed in the Yuba River near Camptonville. With the ability to lift 100–200 gallons of water per minute, they allowed the power of the river to be harnessed. The waterpower could be used to operate mills, pumps, or other equipment. The wheels were also used to divert water into flumes, supplying water for mining. (Both, courtesy of Yuba Feather Museum.)

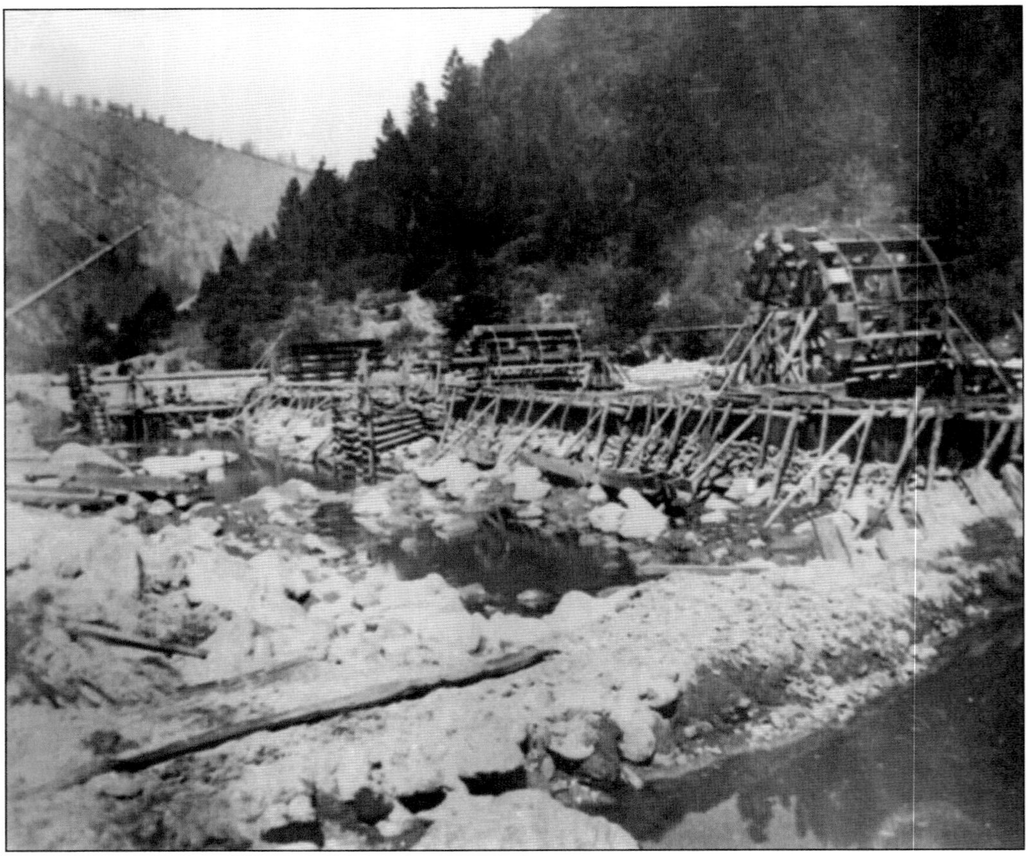

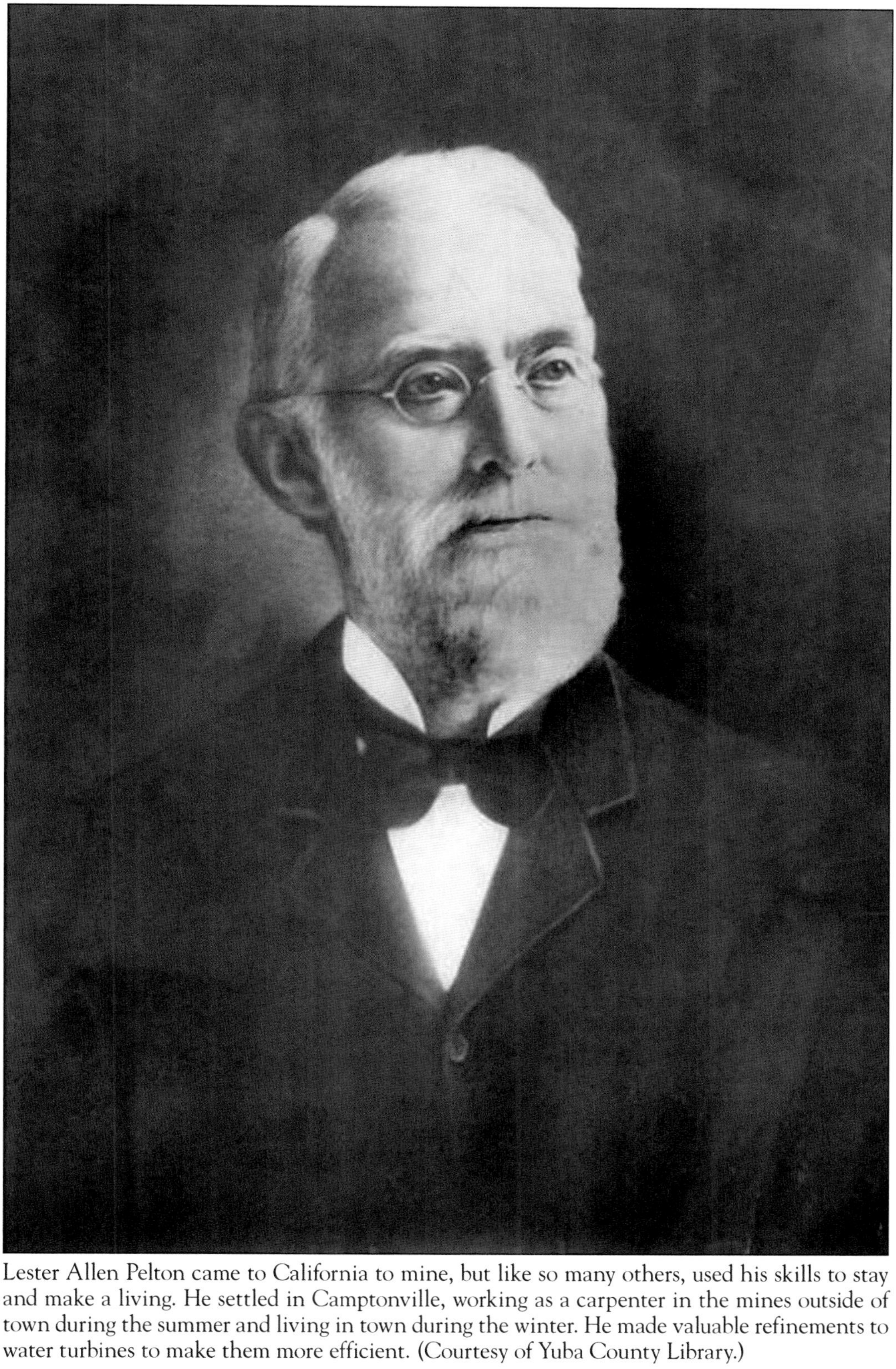

Lester Allen Pelton came to California to mine, but like so many others, used his skills to stay and make a living. He settled in Camptonville, working as a carpenter in the mines outside of town during the summer and living in town during the winter. He made valuable refinements to water turbines to make them more efficient. (Courtesy of Yuba County Library.)

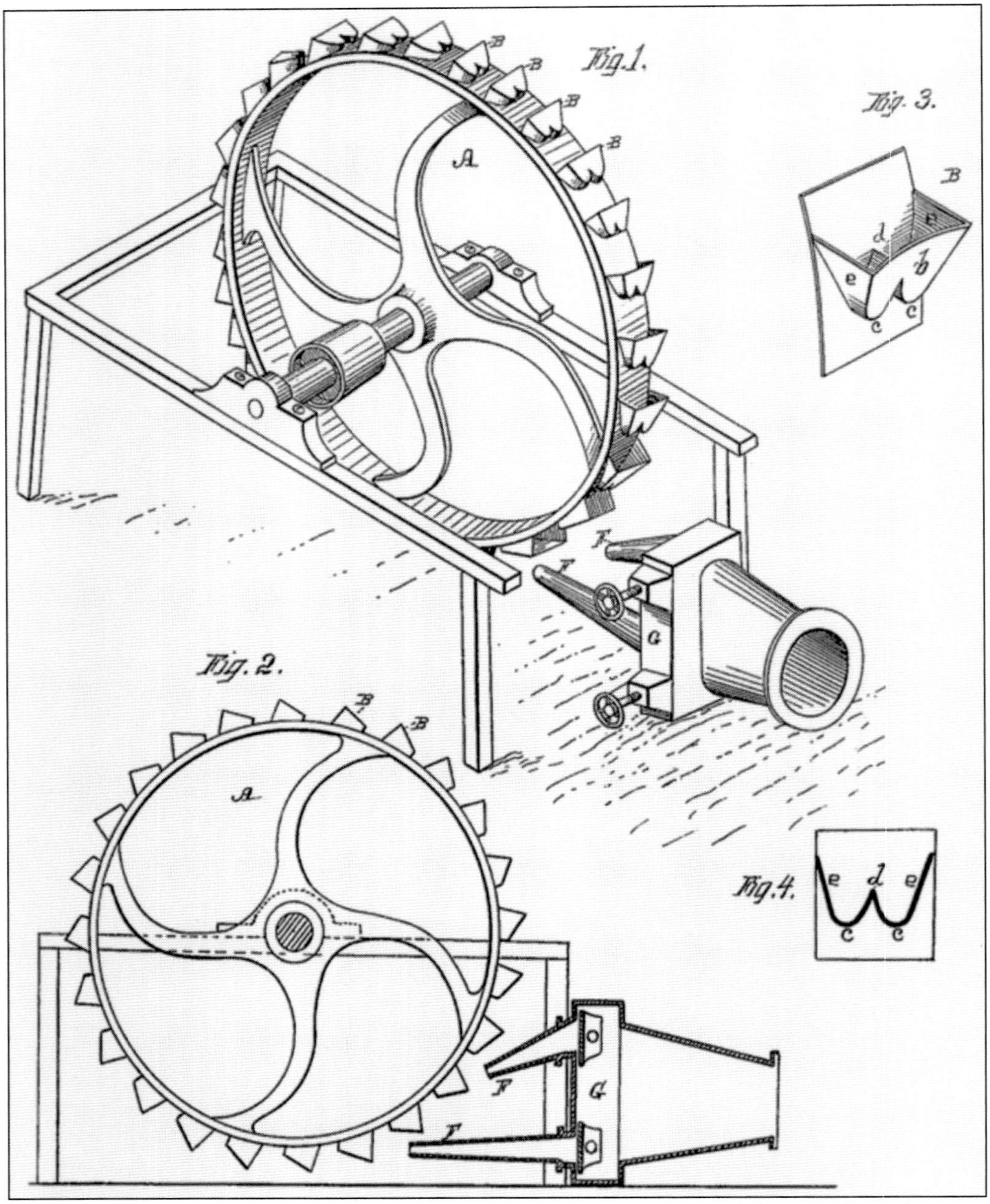

In 1889, Lester Pelton was issued a patent for his improved water wheel design. These drawings detail the innovations that made his design so successful. The improvements did not just benefit mining, but many industries around the world. (Courtesy of Library of Congress.)

Three

There Must Be a Better Way

Hydraulic Mining

When some of the excitement of gold mining wore off, many disillusioned fortune hunters returned to their homes. Those who stayed had a more realistic idea of what it took to extract the gold that remained in the Yuba foothills. The greatest quantities of gold, they were learning, were in the auriferous gravels. This gravel could be found in the extinct channels of ancient rivers, embedded with gold deposits that dated back 50 million years to the Eocene epoch. As the Sierra Nevadas rose, these deposits were enclosed in the foothills and cliffs that surrounded the Yuba River. It was difficult to get to the auriferous gravel deposits because—as the name implies—they were not comprised of solid rock, like granite or quartz, but of gravel that had been cemented together. Conventional mining techniques were not effective here, and a better method for separating the gold from the gravel deposits was needed.

The abundance of water in the region, and the prior, successful diversion of water for power generation, led to the miners diverting even more water into giant, ground-sluicing projects. Ditches were dug and flumes were built to divert vast quantities of water and put it under pressure. The water was fed into hoses or steel pipes and then blasted at the hillsides, washing the ground away into gigantic sluices that caught most of the gold and emptied the waste into the river. The process was known as hydraulic mining, and it became mining's next big thing. The monitor and nozzle, which were produced locally, were perfected and made even more effective. All of this was very expensive of course, and each company was competing with its neighbors for capital investment. Each of them needed to prove the value of their mineral deposits, their ability to get them, and—most importantly—the cost. Geologists and mining engineers soon became experts in great demand.

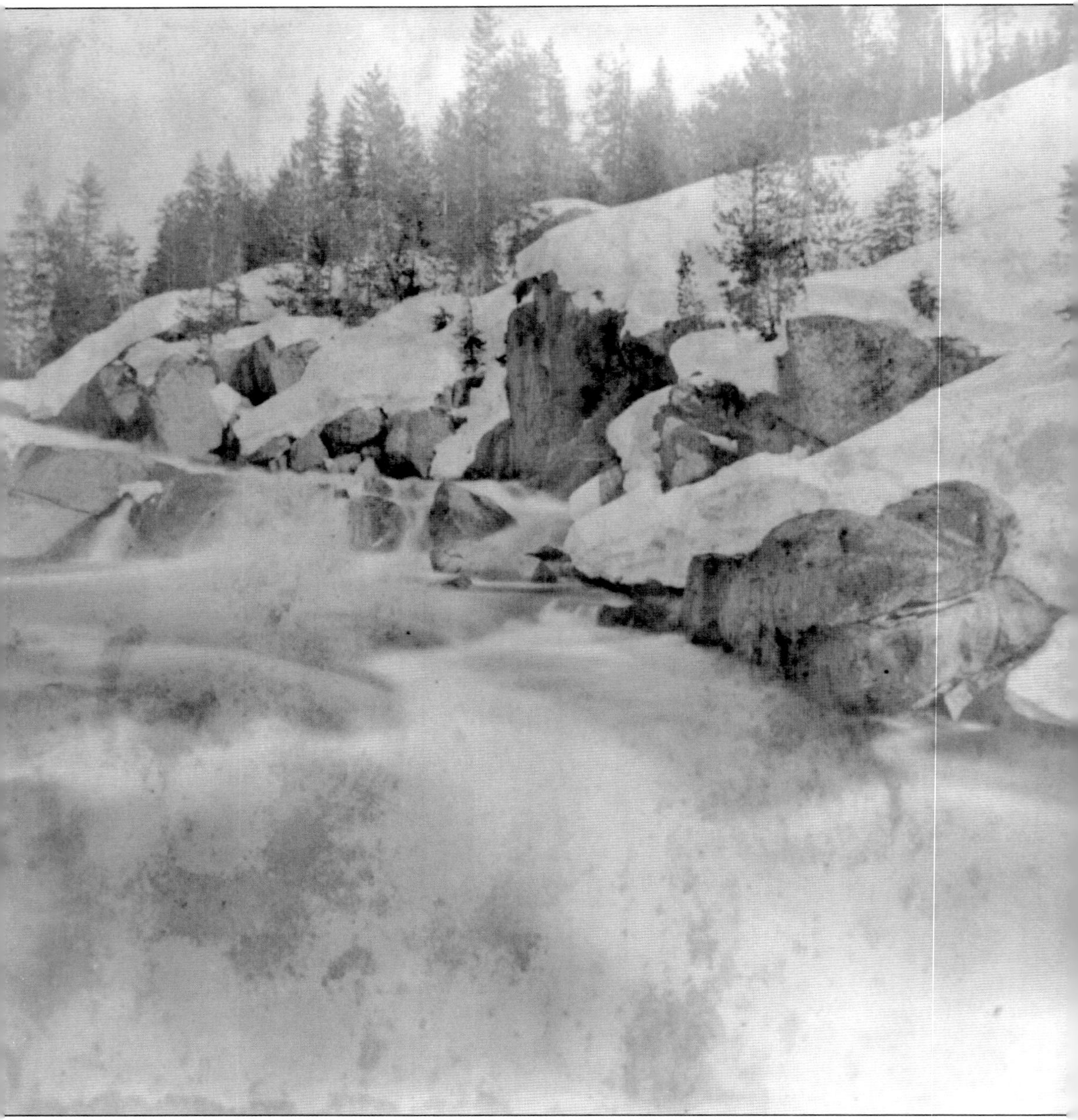

The immense amount of water needed for hydraulic mining came from the headwaters of the Yuba River. The enormous, snow-capped Sierras were the source of the vast quantity of water available in the Yuba watershed. As the snow melted, it traveled down the creeks and ravines into the Yuba and its many tributaries. There were no restrictions or regulations on water use at that time, so whoever was clever enough to engineer a method of harvesting the water was free to do so. (Courtesy of Library of Congress.)

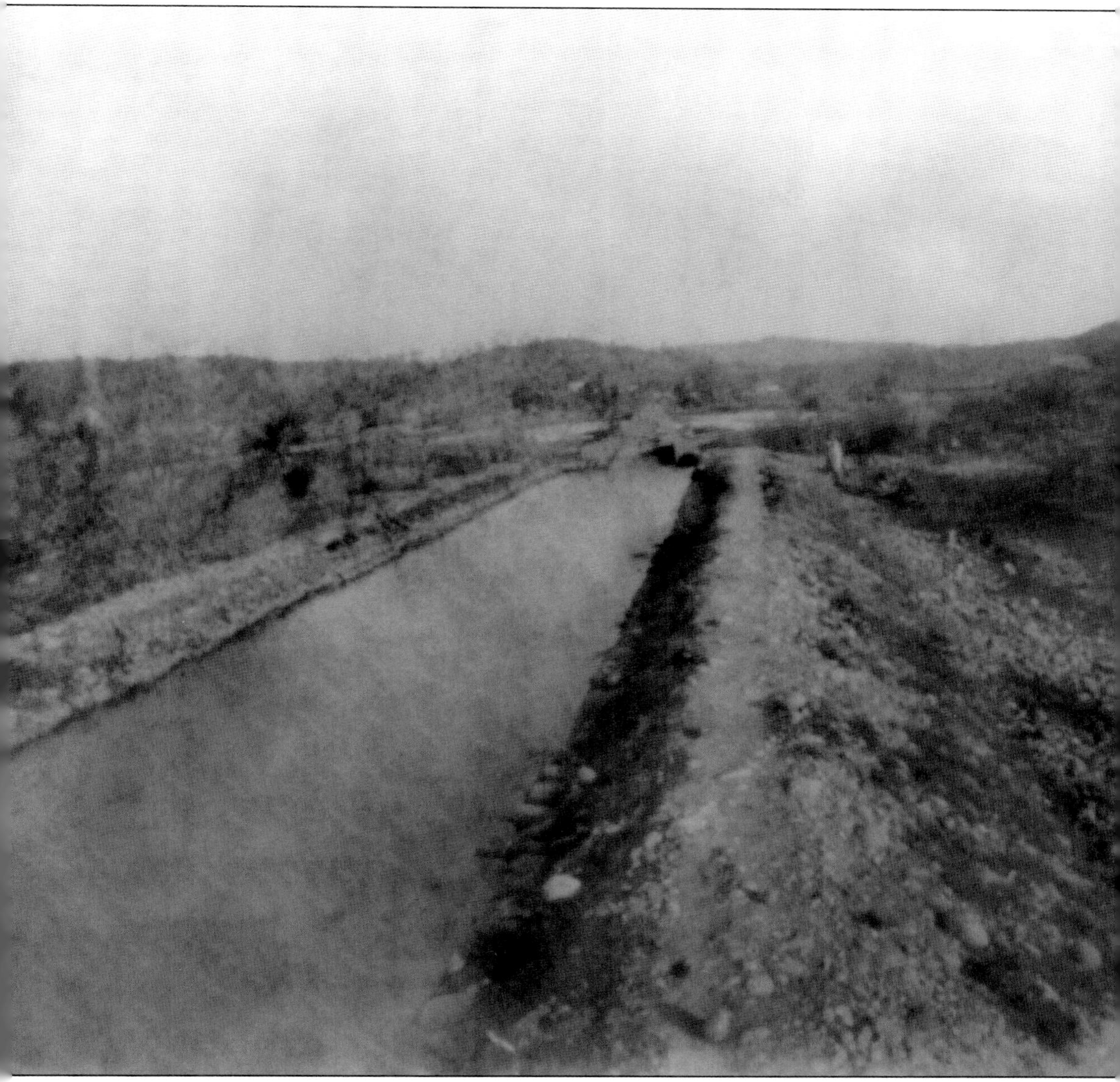

In April 1857, D. Bovyer conceived the idea of building a ditch from Deer Creek reaching 15 miles to Smartsville and Timbuctoo. At the time, this was a questionable endeavor, but it turned out to be successful, so in 1858, James O'Brien built the Boyer ditch, which spanned 25 miles from Deer Creek to Smartsville. The following year, he built the Excelsior Ditch—also known as the China Ditch—from the South Yuba to Smartsville, spanning a distance of about 34 miles. (Courtesy of Library of Congress.)

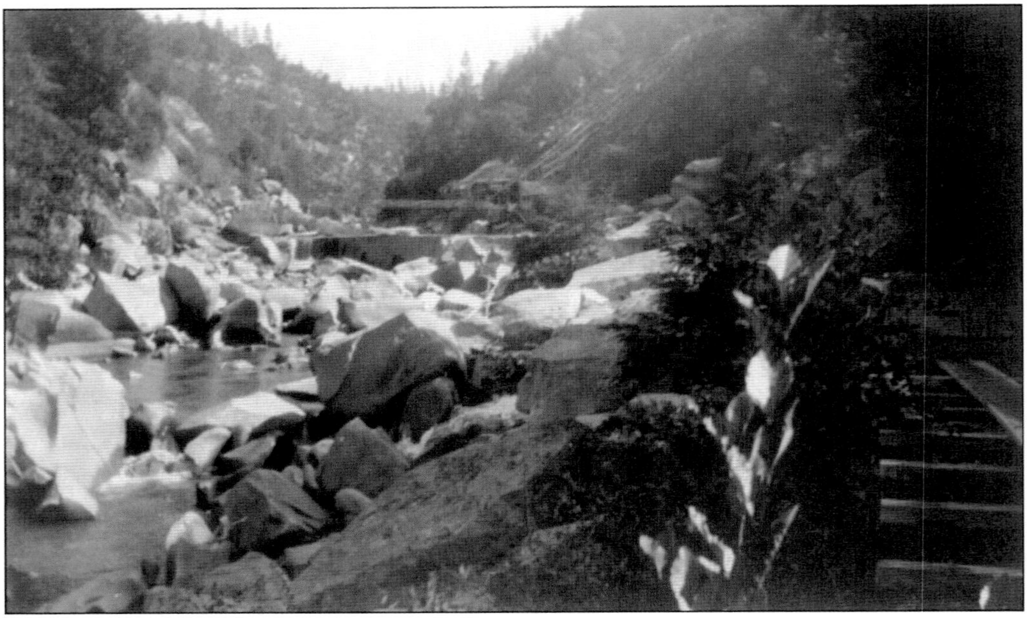

Deer Creek appears to be the point of origin for the canal pictured above. The fast-running creek cuts through the ravine above, moving from the Sierras to its confluence with the Yuba River just below Mooney Flat. The flume below illustrates what an amazing engineering feet this structure was, and the treacherous terrain through which these ditches traveled. (Both, courtesy of the Marple/Hapgood Family.)

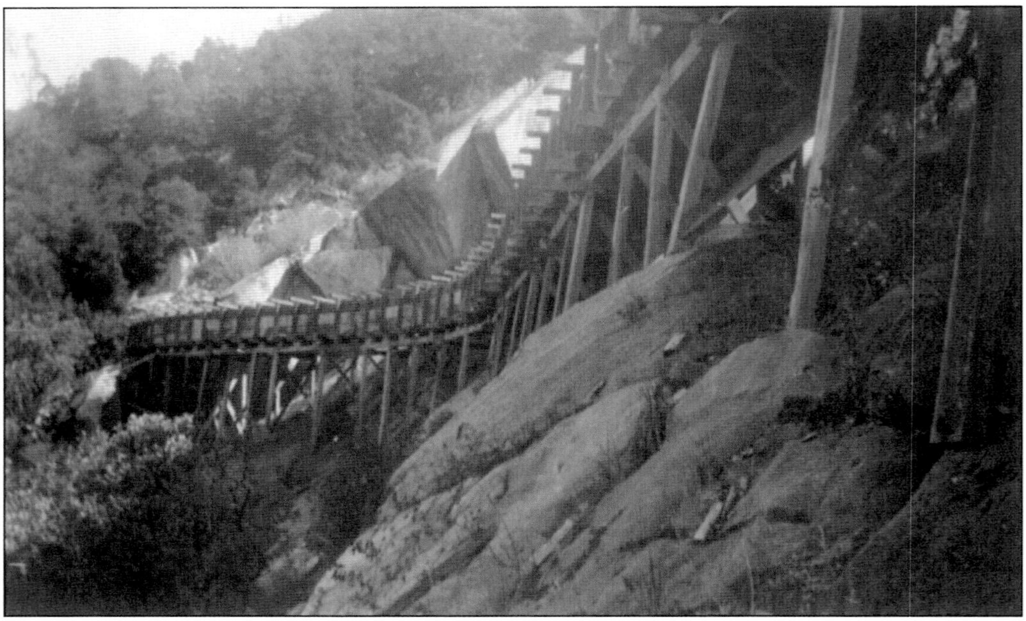

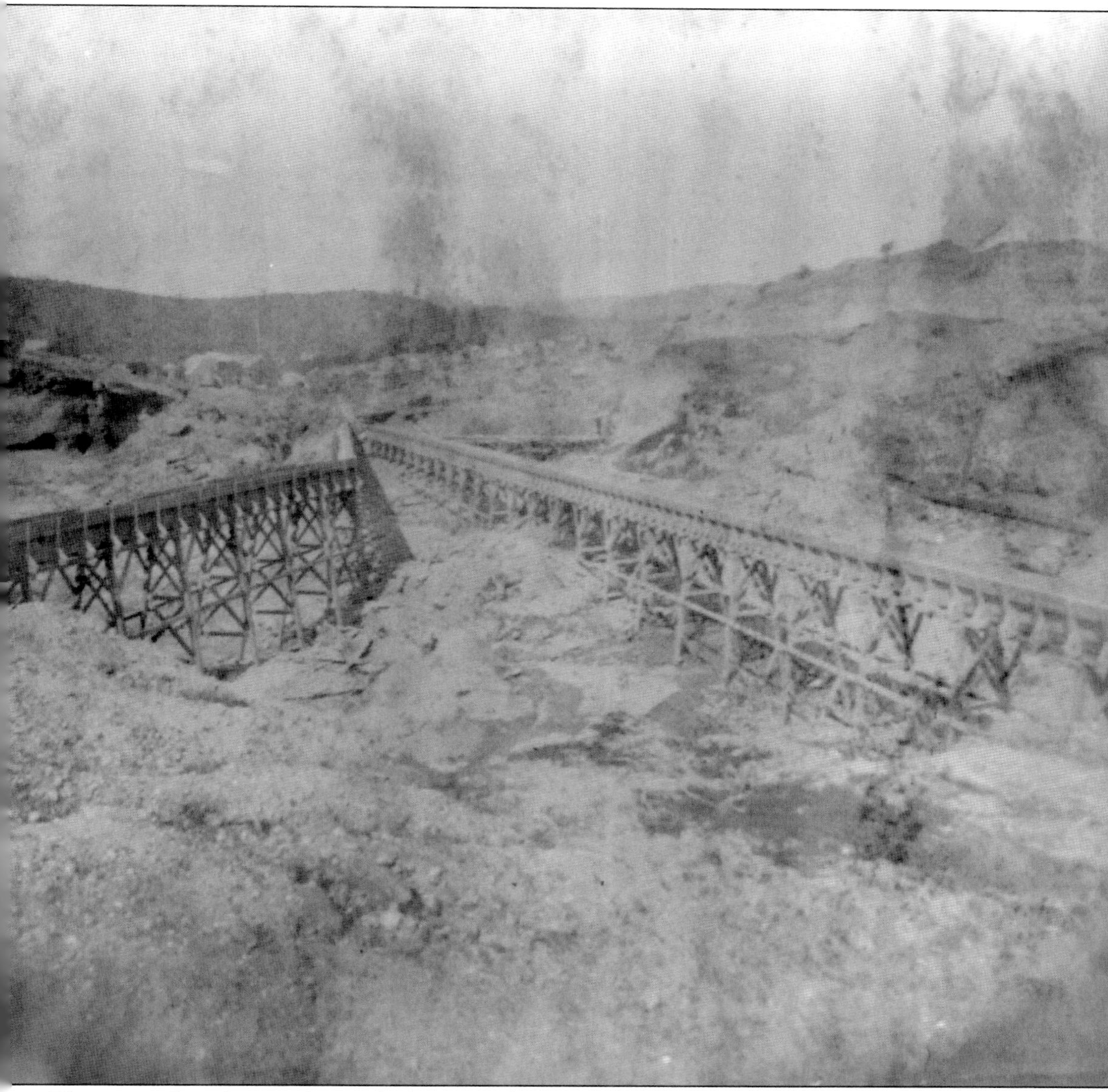

The water that the canals supplied—equivalent to 5,000 inches during the wet season—was conveyed across the ravine between Smartsville and Sucker Flat via a pair of 2,024-foot-long flumes. The largest one, which appears to rise 60–80 feet above the lowest part of the gap, seems an elegant and substantial structure. Another Excelsior Company ditch supplied 4,000 inches and cost about $250,000. In 1861, the Excelsior and Tri-Union ditches combined, forming the Excelsior Canal Company. (Courtesy of Library of Congress.)

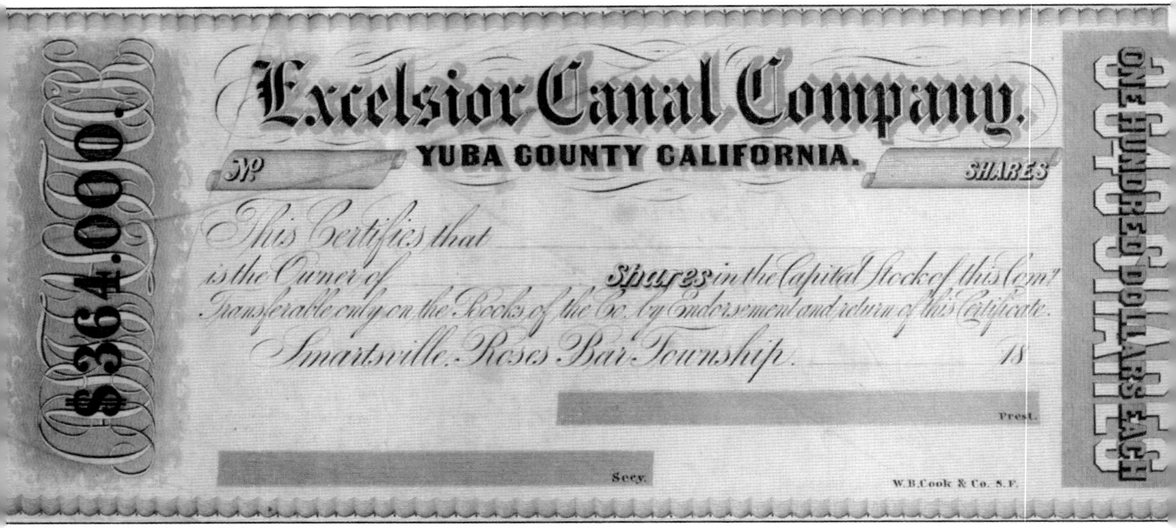

Money was raised to finance these extravagant canal projects by selling shares. For example, this Excelsior Canal Company stock certificate would have been filled in with the investor's name and the quantity of shares owned—bought at a rate of $100 per share. The mines also sold stock to finance their improvements. If projects were running low on funds, there would be an assessment on the shareholders for more money. If the shareholder couldn't pay the assessment, they would lose their stock and their investment. (Courtesy of California Society of Pioneers.)

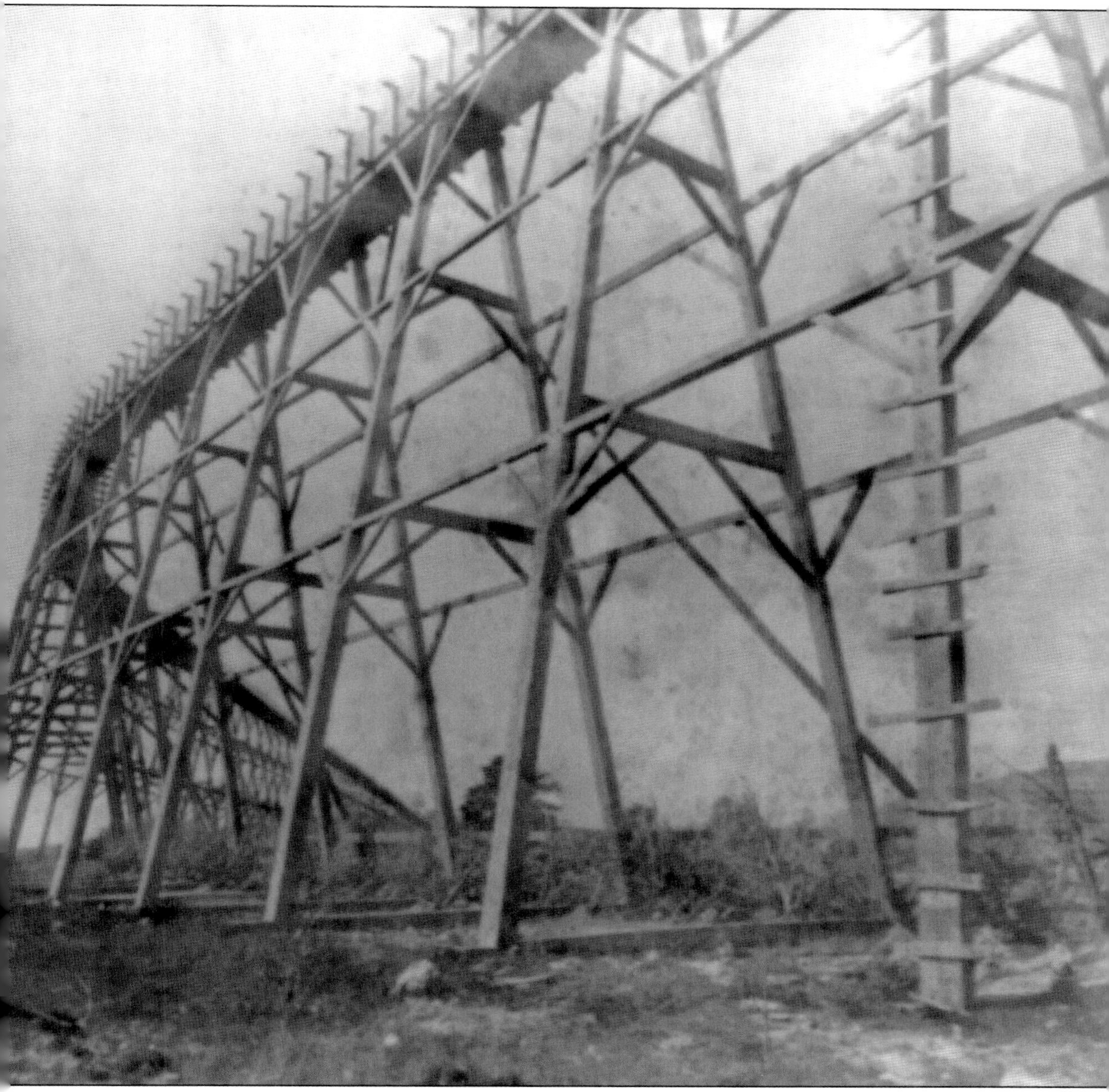

On January 21, 1858, high winds blew down the Excelsior Company's aqueduct at Timbuctoo. This great work, which carried a high cost in both money and labor, had just been completed and had not yet begun to perform the duty for which it was erected. The estimated loss to the Excelsior Company was at least $40,000. (Courtesy of Library of Congress.)

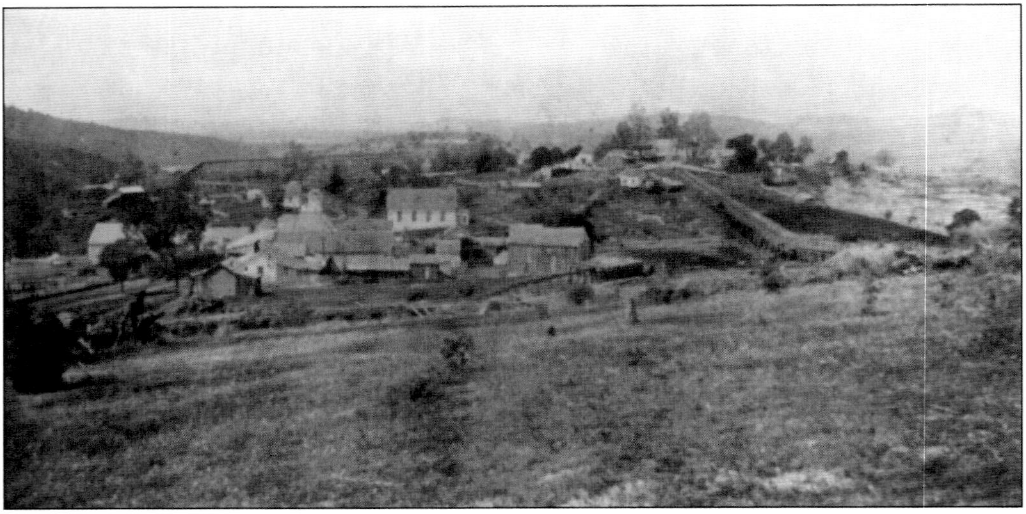

In this photograph, the flumes supplying water to Timbuctoo can be seen at the right going up Temperance Hill, behind Smartsville, and into Timbuctoo on the left. This demonstrates how intertwined the mining industry was with the towns that supported it. (Courtesy of Library of Congress.)

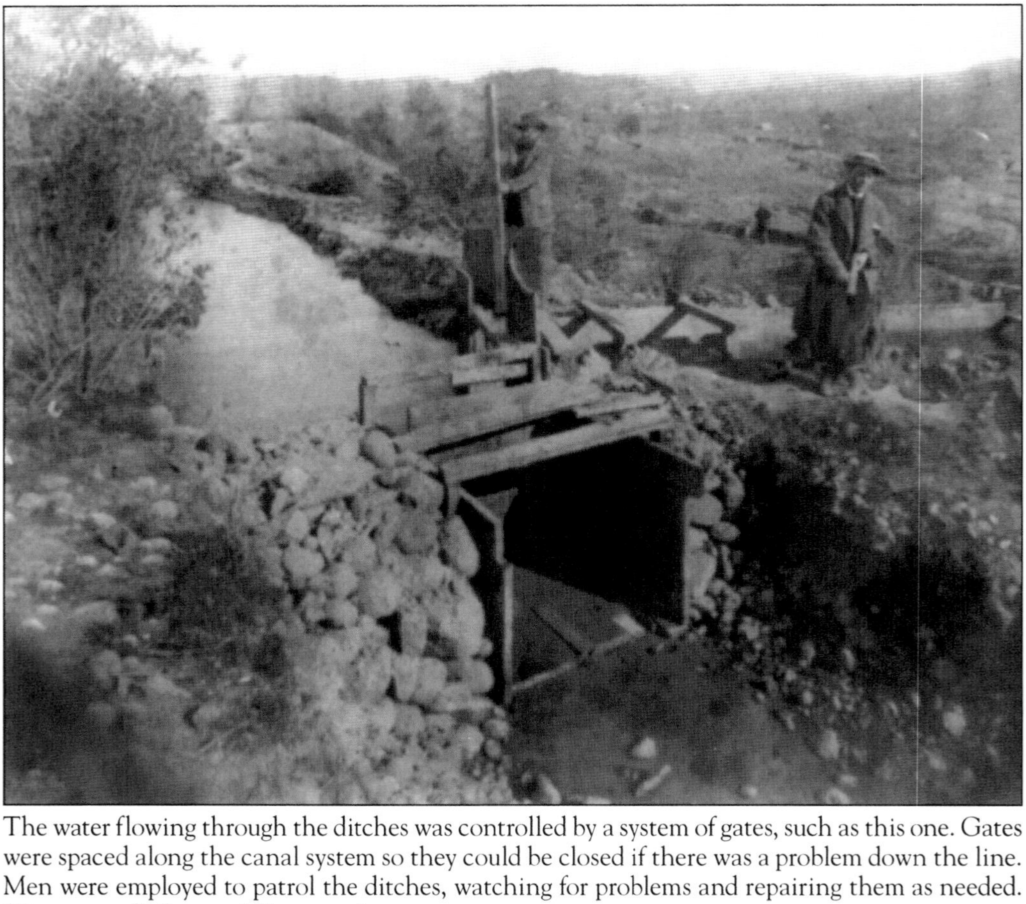

The water flowing through the ditches was controlled by a system of gates, such as this one. Gates were spaced along the canal system so they could be closed if there was a problem down the line. Men were employed to patrol the ditches, watching for problems and repairing them as needed. (Courtesy of Library of Congress.)

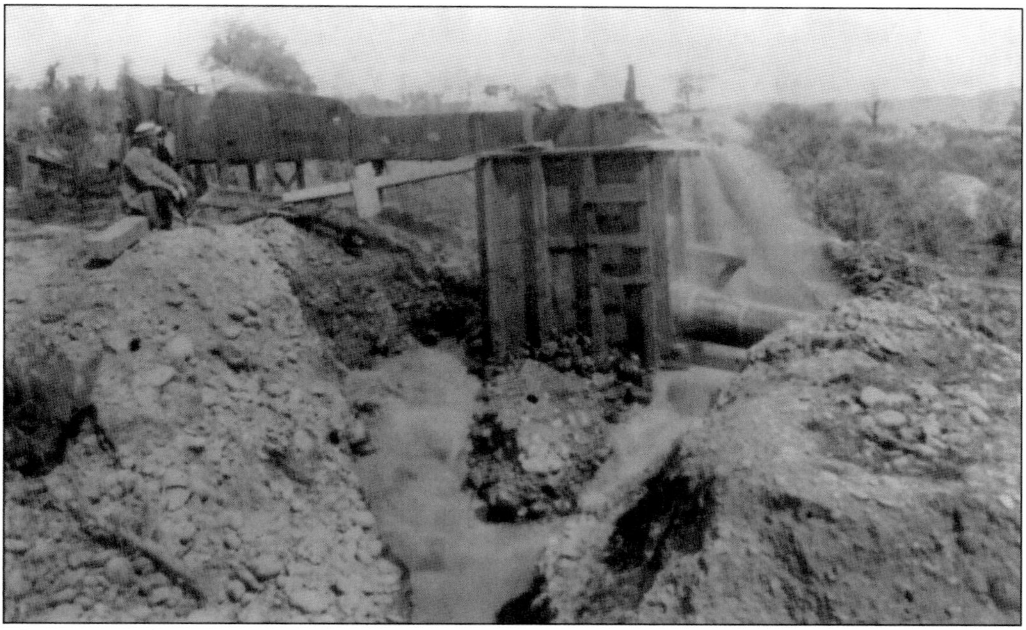

This pressure box was part of the process of pressurizing canal water—serving as the transition point between the ditch and the supply pipes. From this point, the water was sold to the mines by "the miner's inch." This unit of measurement refers to a one-inch pipe kept under at least six inches of pressure and putting out 11.22 gallons of water per minute. The pressure box was placed deep enough to keep the head of the pipe well under pressure. (Courtesy of Library of Congress.)

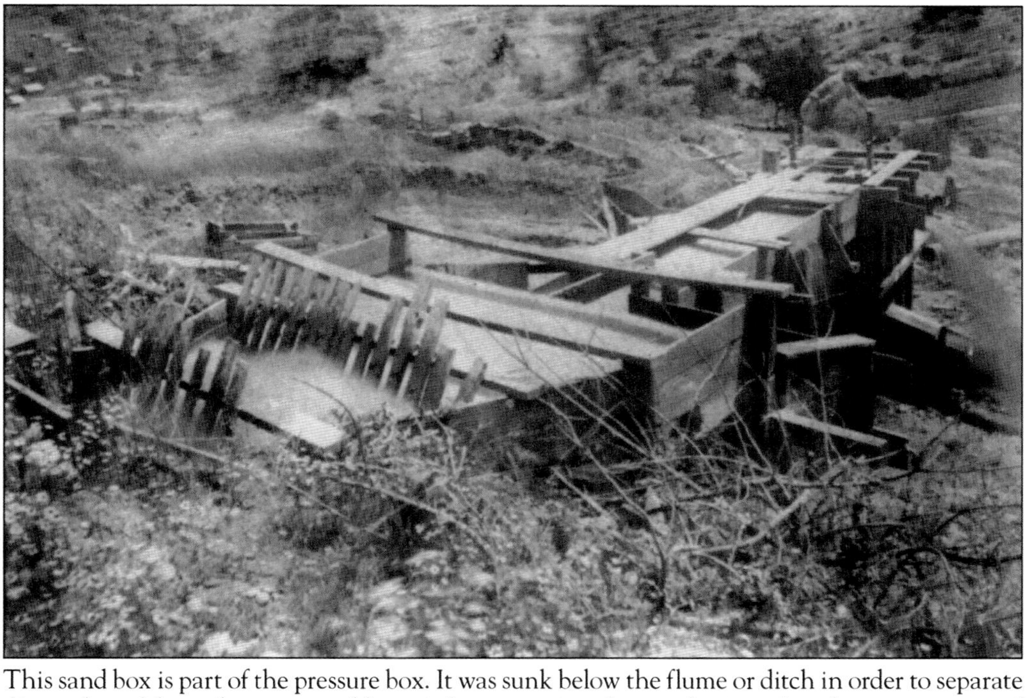

This sand box is part of the pressure box. It was sunk below the flume or ditch in order to separate dirt and sand from the water and keep it from entering the supply pipe. Another component of this process was a grating that caught floating debris like leaves or twigs. A watchman would be kept on site at the sand box to empty it via a side gate. (Courtesy of Library of Congress.)

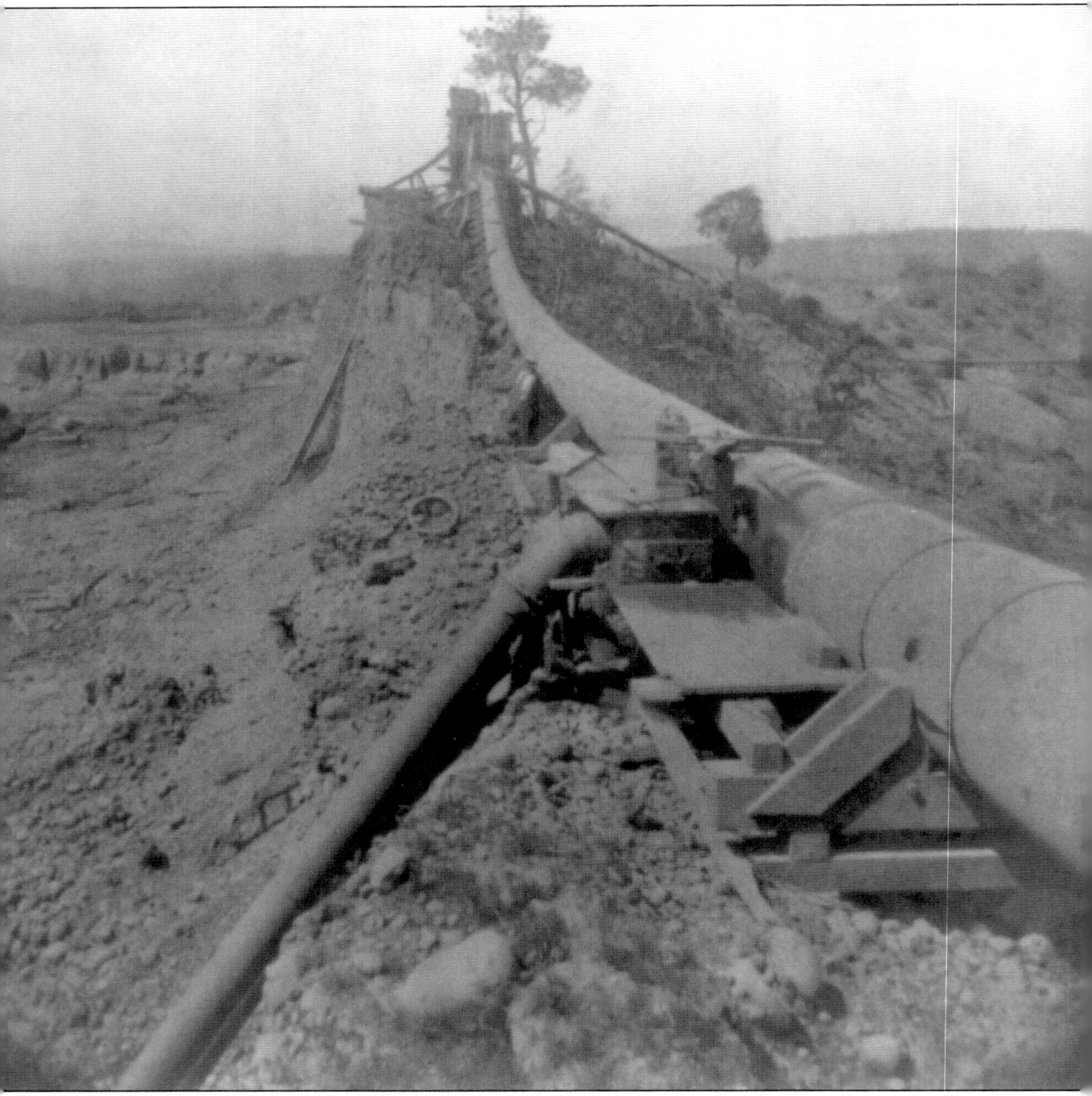

This appears to be the backside of a pressure box at the top of an incline. This configuration of the decline of the pipe coming from the pressure box and the narrowing of the outlet for the water created more pressure, while delivering the water to the spot where it was needed. (Courtesy of Library of Congress.)

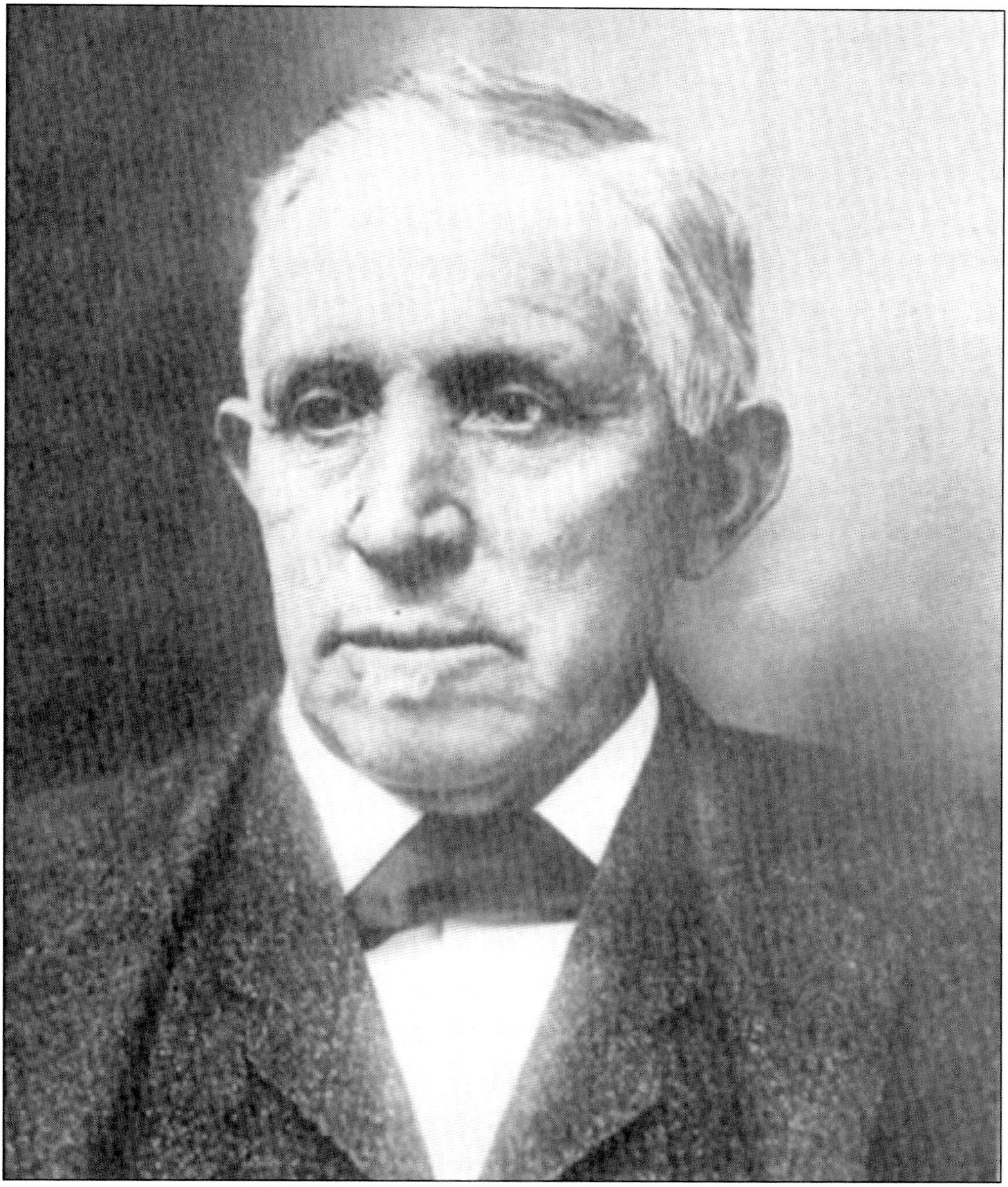

James O'Brien arrived in California in 1853 and began his mining career at Barton's Bar. After a stint building roads in Yolo, he was contracted to build many of the area ditches, for which he employed Chinese laborers. This acquainted him with the cost of the improvements needed to compete in the hydraulic mining business. In the fall of 1859, he bought mining property near Smartsville. In 1865, he commissioned a report from Prof. William Ashburner on the feasibility of profits for his mining properties. The purpose of the report was to encourage investment in his promising claim. In 1868, he incorporated with investors Ashburner, Walker, Baker, and Hague of San Francisco for the building of the Pactolus Tunnel at a cost of $80,000. O'Brien owned 50 percent of the shares, was made superintendent of the mine, and ran it for many years. It was later consolidated into the Excelsior Water & Mining Company, and even with the added responsibility, he continued as superintendant of that company for another four years. The property was eventually sold to Eastern capitalists. (Courtesy of YubaRoots.)

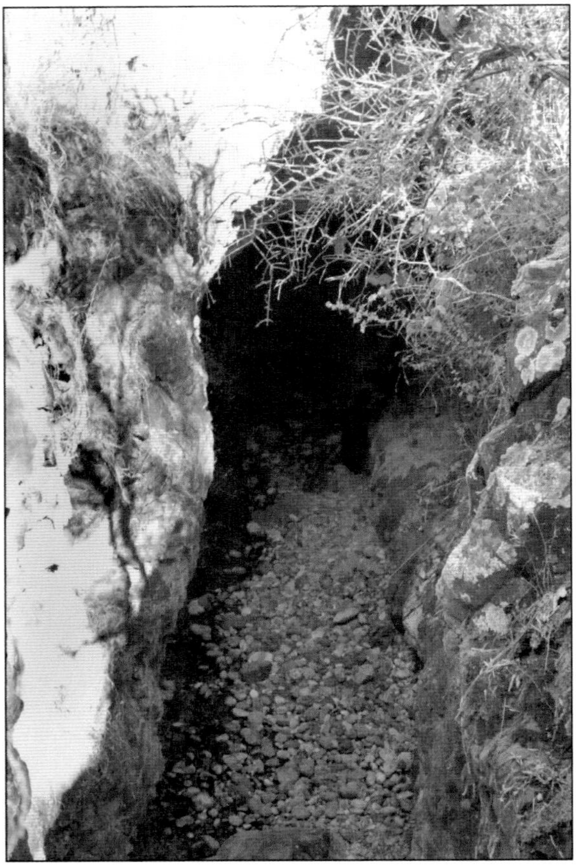

The tunnels were the most expensive part of hydraulic mining—ranging from 500 to 4,000 feet in length and costing from $12 to $50 per foot to dig. The tunnels were cut through the bedrock from the river and the tailings were discharged into an opening at the bottom of the mine. They are large enough for a man to walk through, and their bottoms are watertight and paved with rock. Pictured at left is the opening of the Blue Point Tunnel, which runs underground towards the Yuba River for nearly 10 miles. Below is the exit of the same tunnel. Visible are the blocks that narrow the water flow for entry into the sluice flume boxes that would traditionally be attached here. (Both, courtesy of Brian Bisnett.)

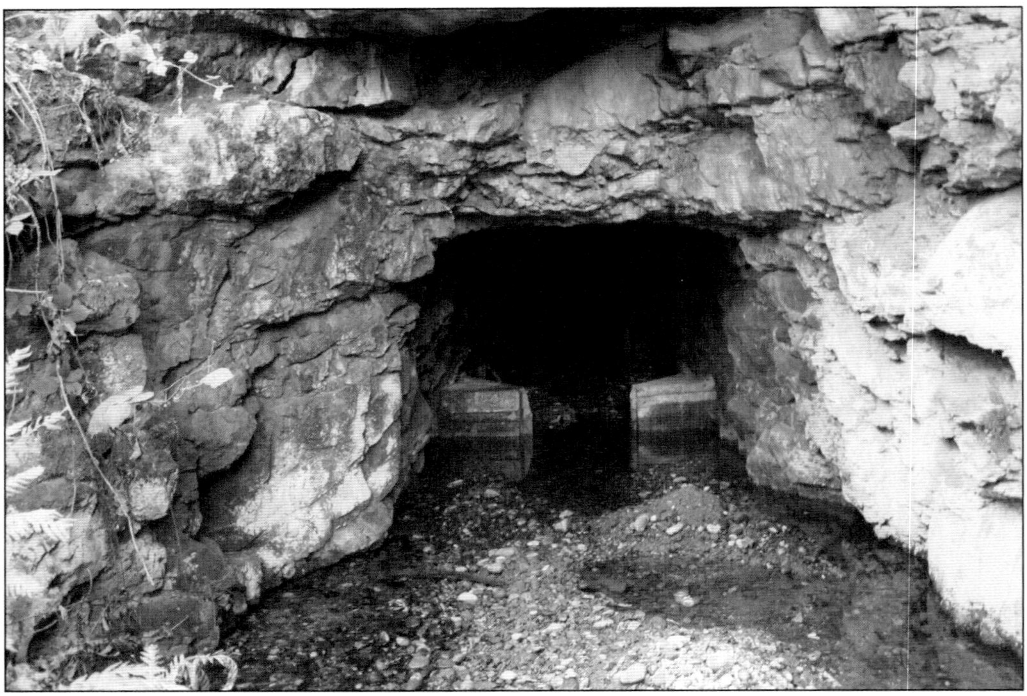

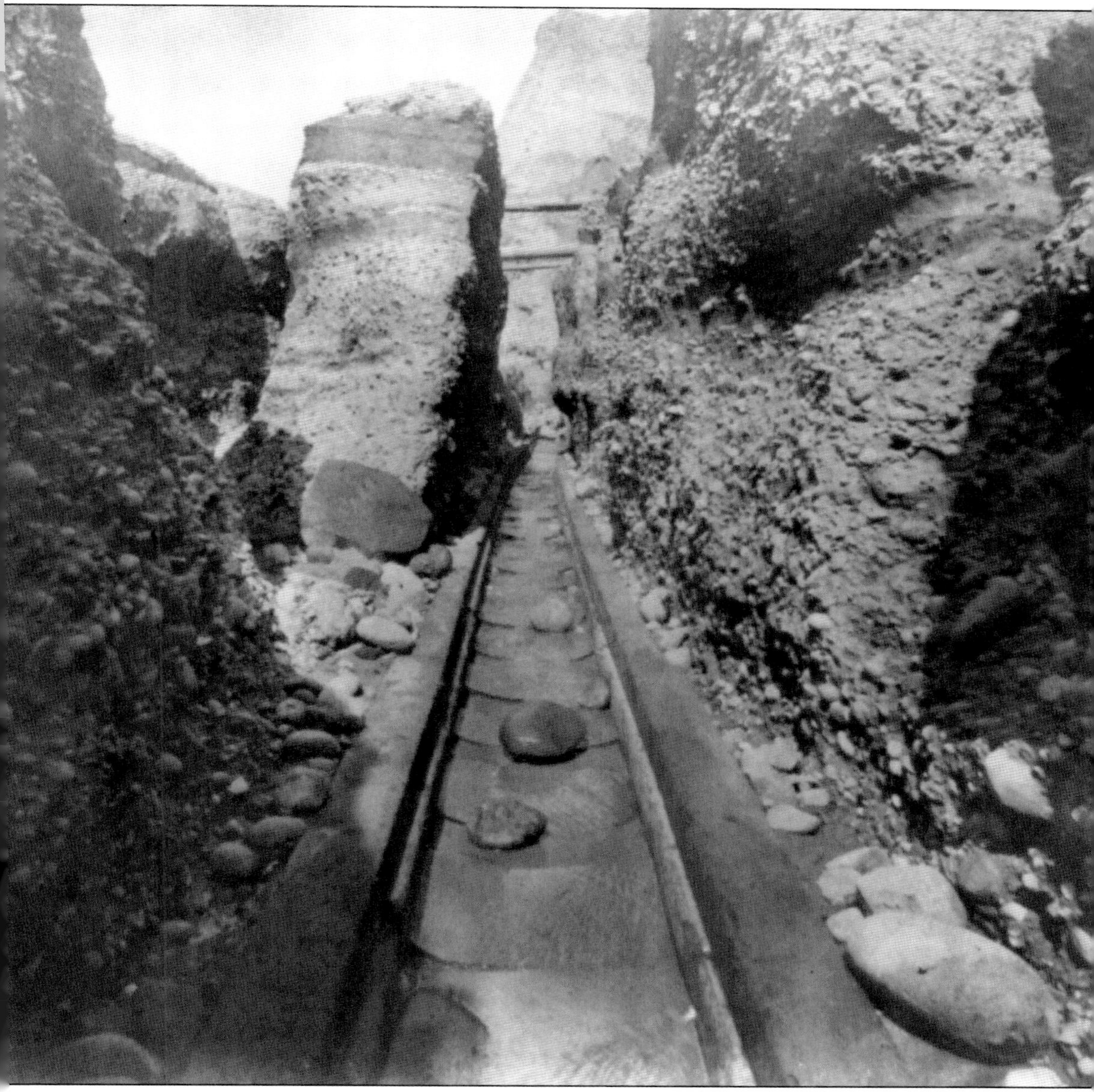

The Blue Gravel Mine in Smartsville was said to be the richest in the state. Its success was due to a 14,000-foot tunnel that ran from the river to the upper portion of the lower stratum—called blue gravel because of its distinctive color. Blocks on the inside of the sluice are shown here. These blocks are approximately 16 inches on a side and 8–10 inches thick. In this arrangement, the blocks are in a single row, but in the sluice boxes, there may be a double row of blocks. Also visible are the crevices around the edges of the blocks, where the gold was trapped before the water and mud were discharged. (Courtesy of Library of Congress.)

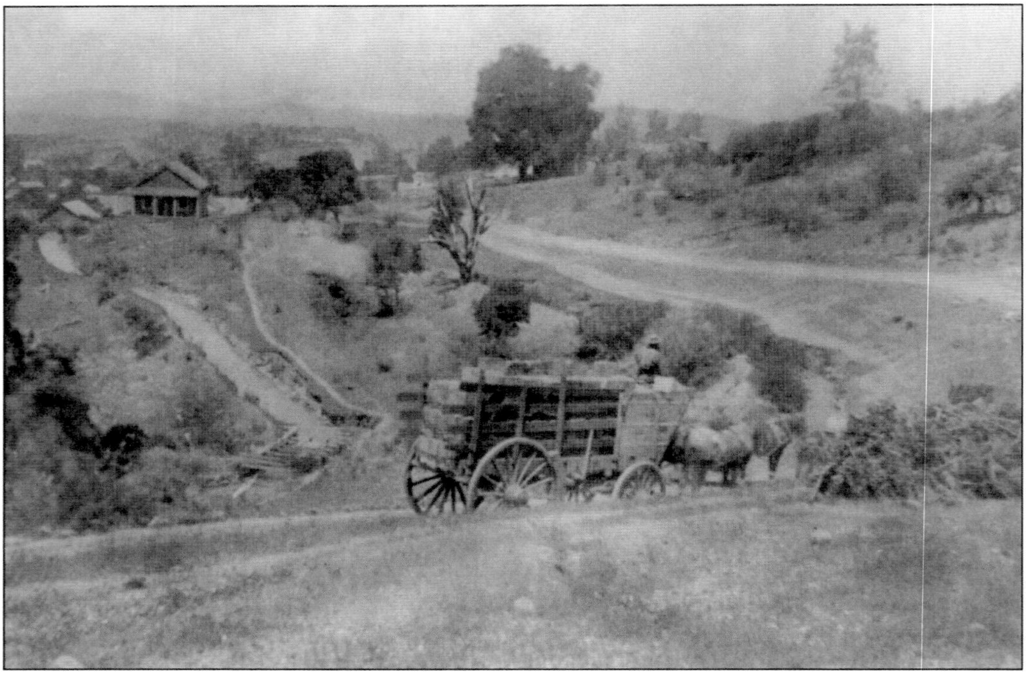

This wagon hauled sluice blocks for the Blue Gravel Mine. Notice the size of the blocks. They are much thicker than they appear when seen in place in the sluice. The blocks were laid closely together so that they wouldn't become loose. Next, the quicksilver was scattered in the sluice. When the tailings were washed down the sluice, the quicksilver attracted the gold and bonded with it to become an amalgam. If a paving block came loose, it could knock the flume loose and it could be washed away downriver. The length of the tail sluices could be shortened or lengthened as needed. The one below appears to be quite long. (Both, courtesy of Library of Congress.)

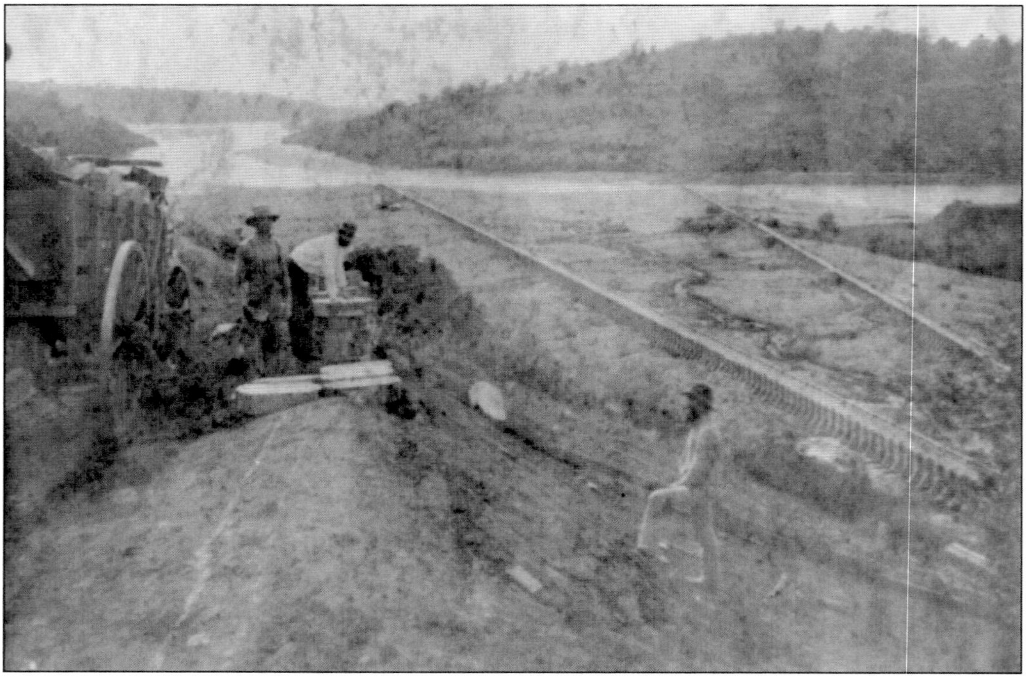

This photograph of the Palm Claim in Timbuctoo shows a larger pipe in the foreground running across the sluice, and smaller pipes winding their way to the background, where the miners are washing the hill down into the sluice at center. (Courtesy of Library of Congress.)

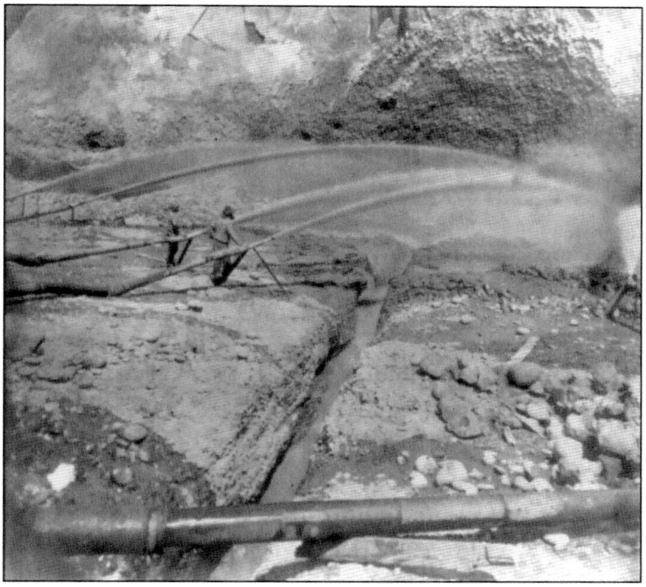

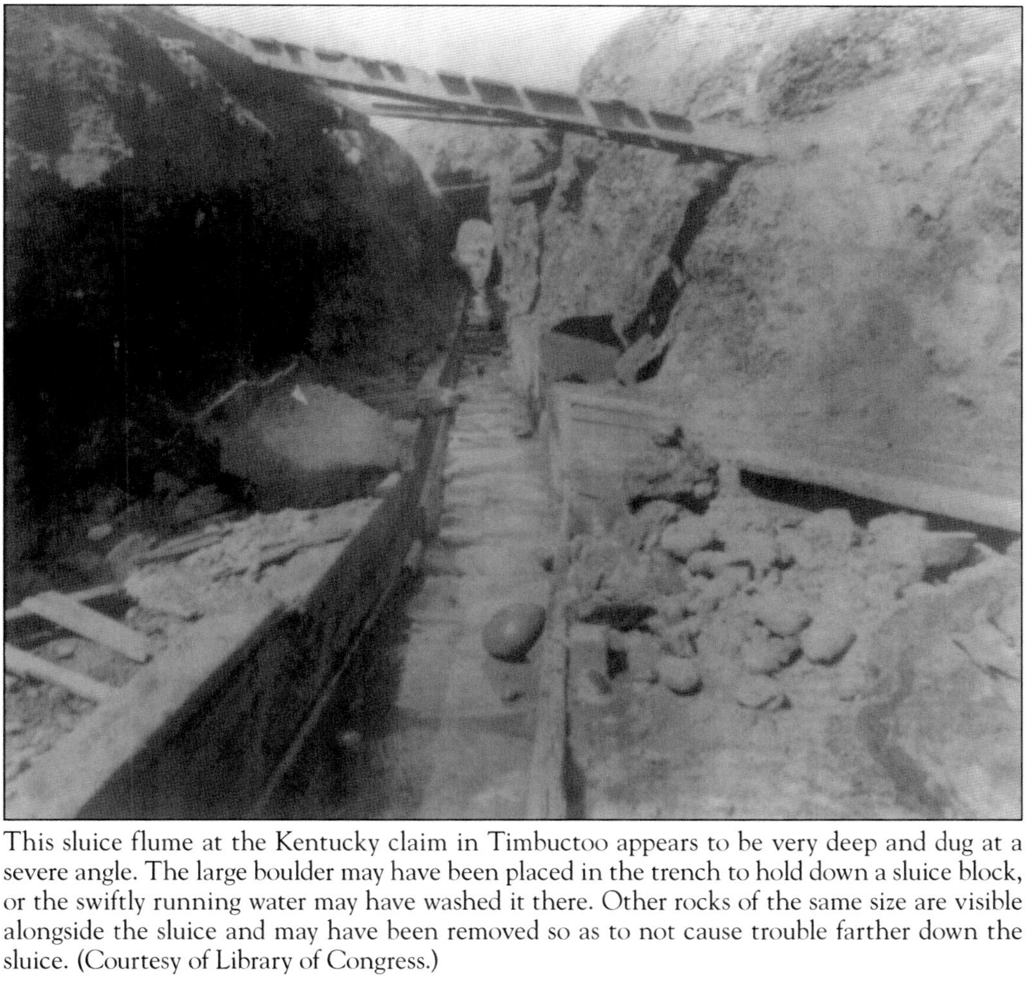

This sluice flume at the Kentucky claim in Timbuctoo appears to be very deep and dug at a severe angle. The large boulder may have been placed in the trench to hold down a sluice block, or the swiftly running water may have washed it there. Other rocks of the same size are visible alongside the sluice and may have been removed so as to not cause trouble farther down the sluice. (Courtesy of Library of Congress.)

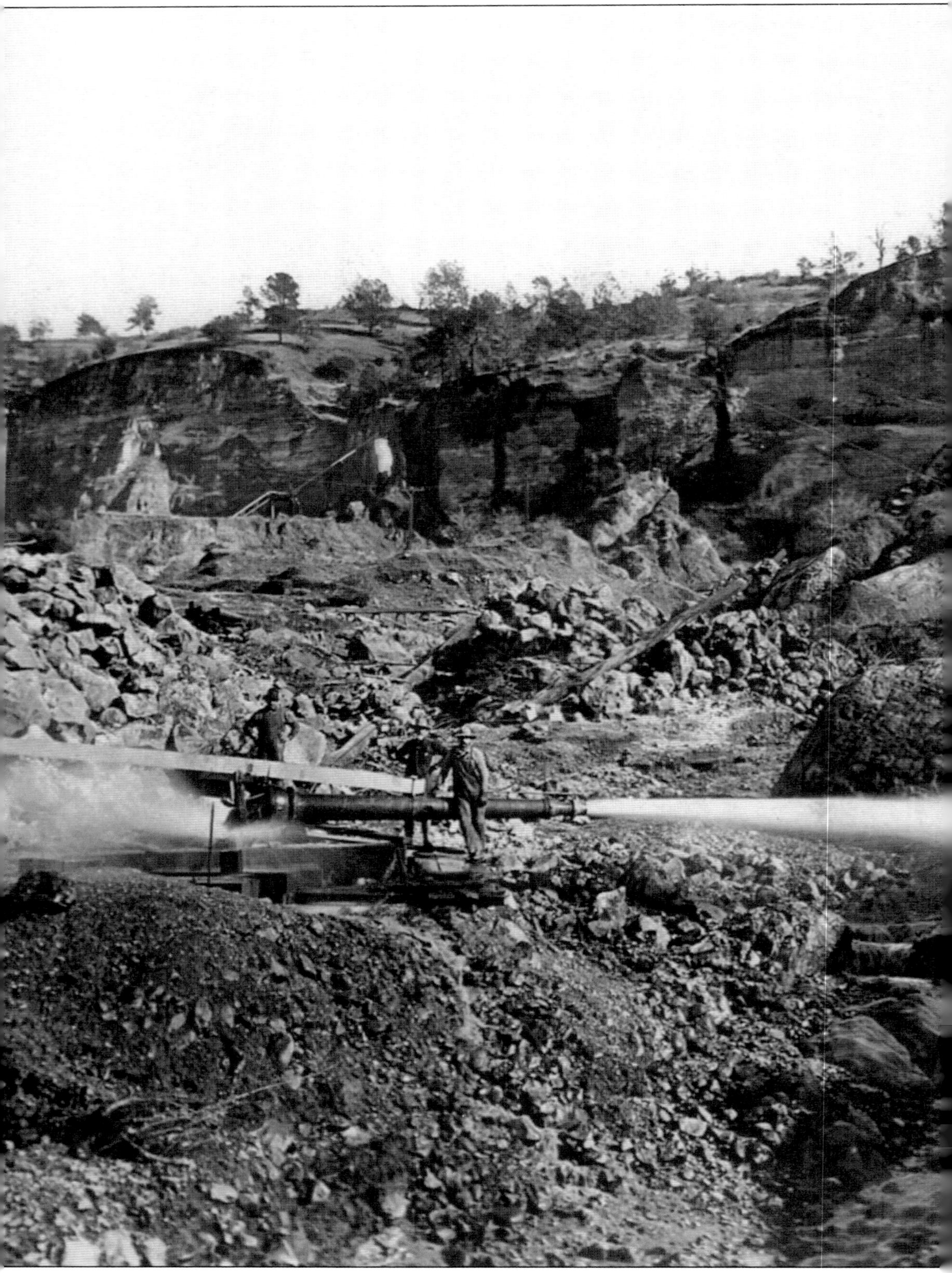

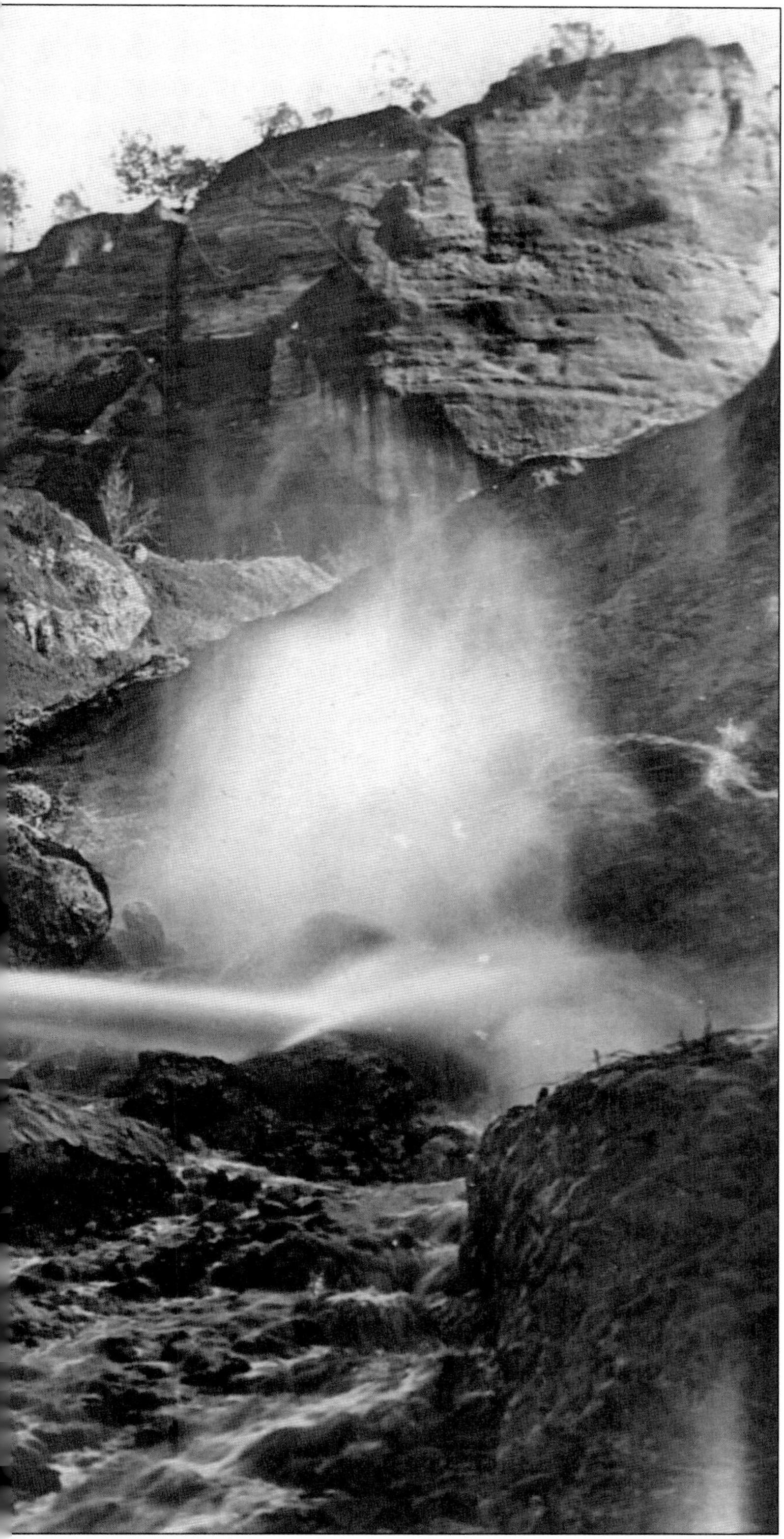

This photograph of the Blue Point Mine in Sucker Flat in action gives an excellent look at the operation of the hydraulic monitor. The monitor resembles a cannon with a long, tapered barrel, flanged at the back and bolted to the pipe. Under the barrel, a ball socket allows the nozzle to be pointed in any direction. The monitor rests atop a timber with a large wooden box at its back end. This balance box was filled with rocks to balance the weight of the nozzle, allowing a miner to easily maneuver the powerful equipment by adding or removing rocks. (Courtesy of the Marple/Hapgood Family.)

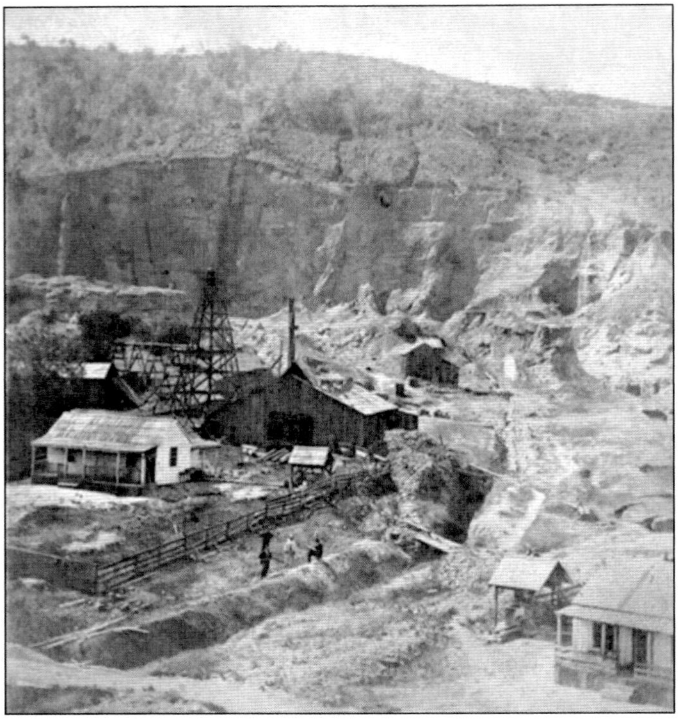

At left, the town of Sucker Flat is shown as a thriving community with family homes, stores, saloons, a bunkhouse, and the headquarters of Carry's Mining. Many of the townsfolk were Irish immigrants who first fled the potato famine of their homeland and then the oppressive conditions on the East Coast. It appears the miners are washing a hill down around the buildings. Men in suits, possibly investors, appear to be observing the operations. Below, hydraulic monitors are seen blasting away at a hill near Sucker Flat, visible up on the rim of the ravine. (Both, courtesy of Library of Congress.)

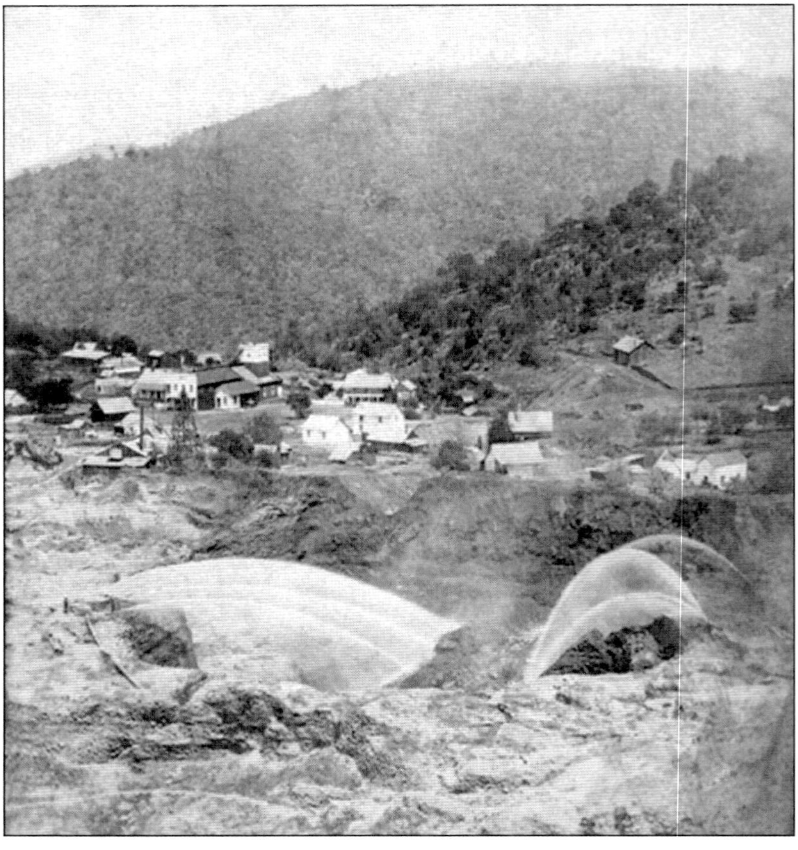

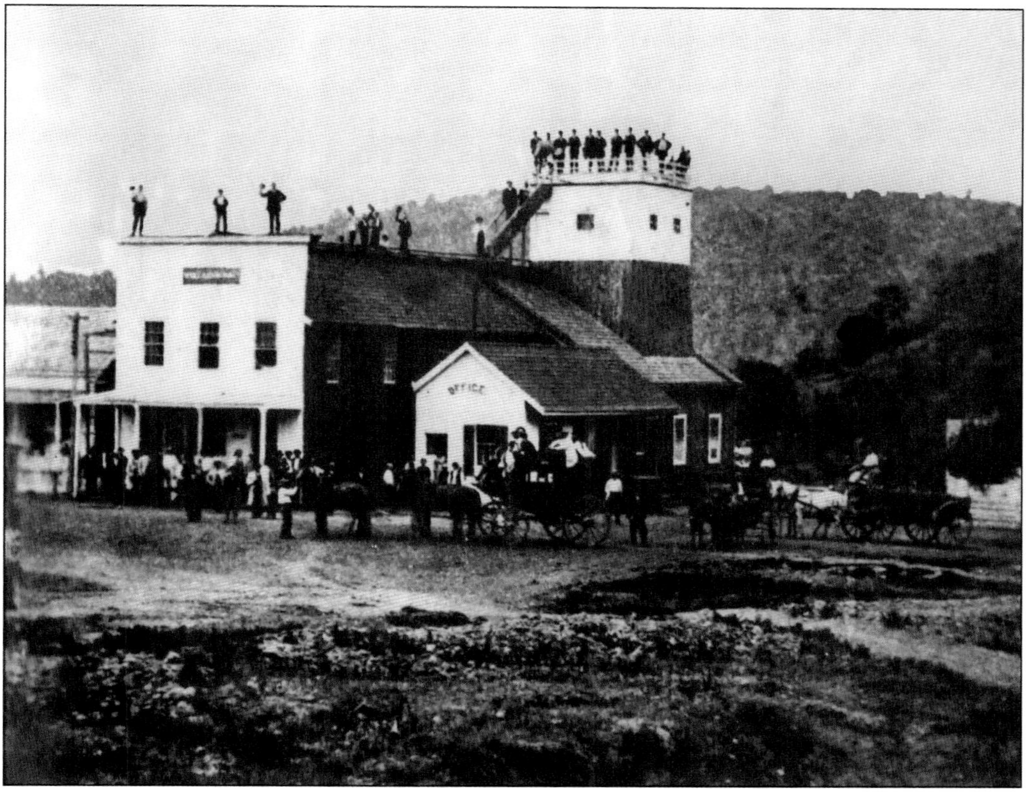

Payday in Sucker Flat was a major event. The above photograph shows Carry's Complex, which consisted of offices, a bunkhouse and cookhouse for the single miners, a store, and a saloon. The miners have all assembled to collect their pay and have a celebration. The large tunnels and sluices were cleaned of gold as needed, but the miners only did a thorough cleaning three times a year, as the entire operation had to be shut down. When that happened, the miners were paid and had a day off. At right, the paymaster poses in front of the payroll safe, stuffed with cash. (Above, courtesy of Bill Peardon; right, courtesy of California State University, Chico.)

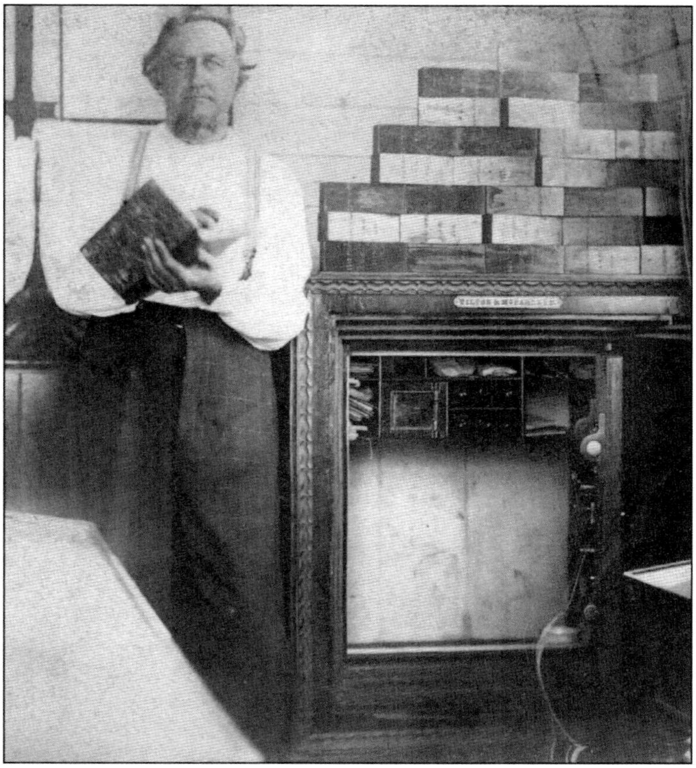

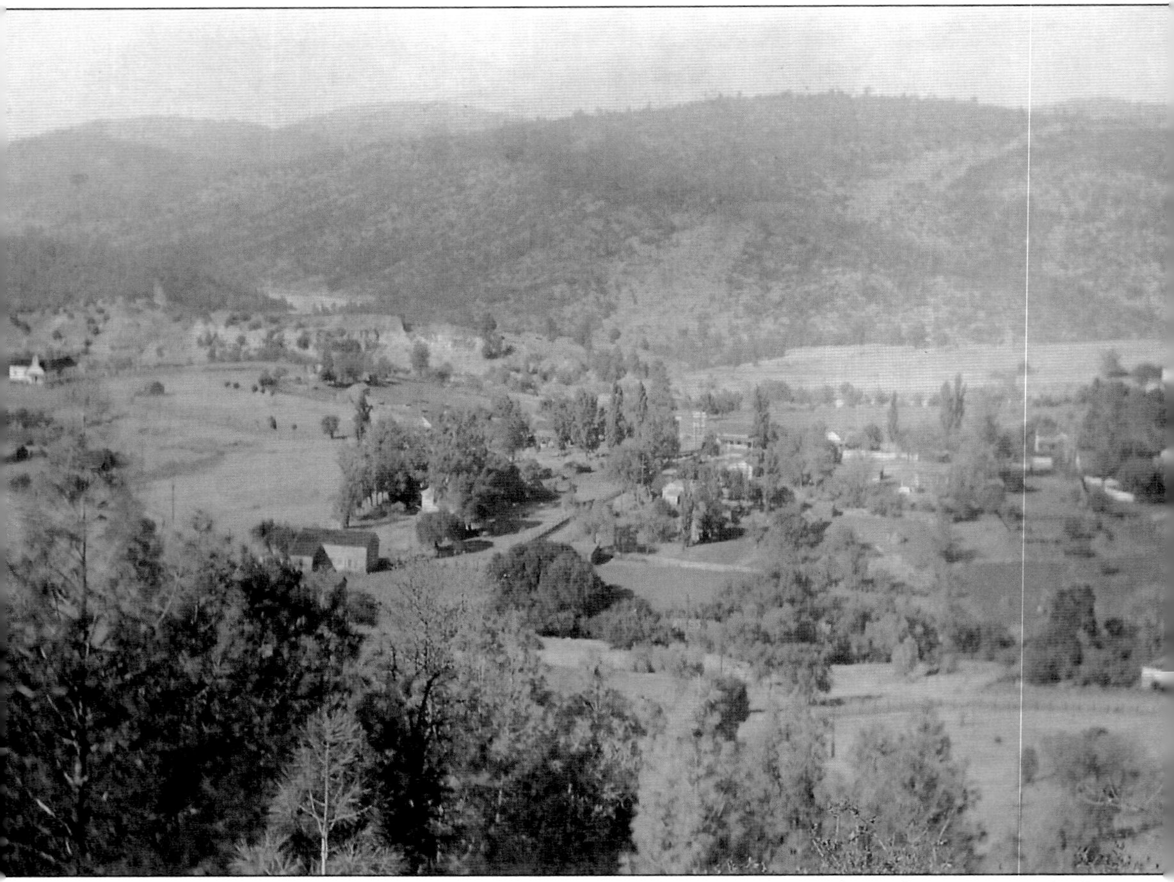

Smartsville was the town that grew out of hydraulic mining. As the tailings encroached on Rose's Bar and began flooding it, several buildings were moved by being put on logs and hauled up Temperance Hill with mule teams. The Masonic lodge was among these buildings, as was the school. As hydraulic mining continued to be used in Timbuctoo and Sucker Flat, most of the

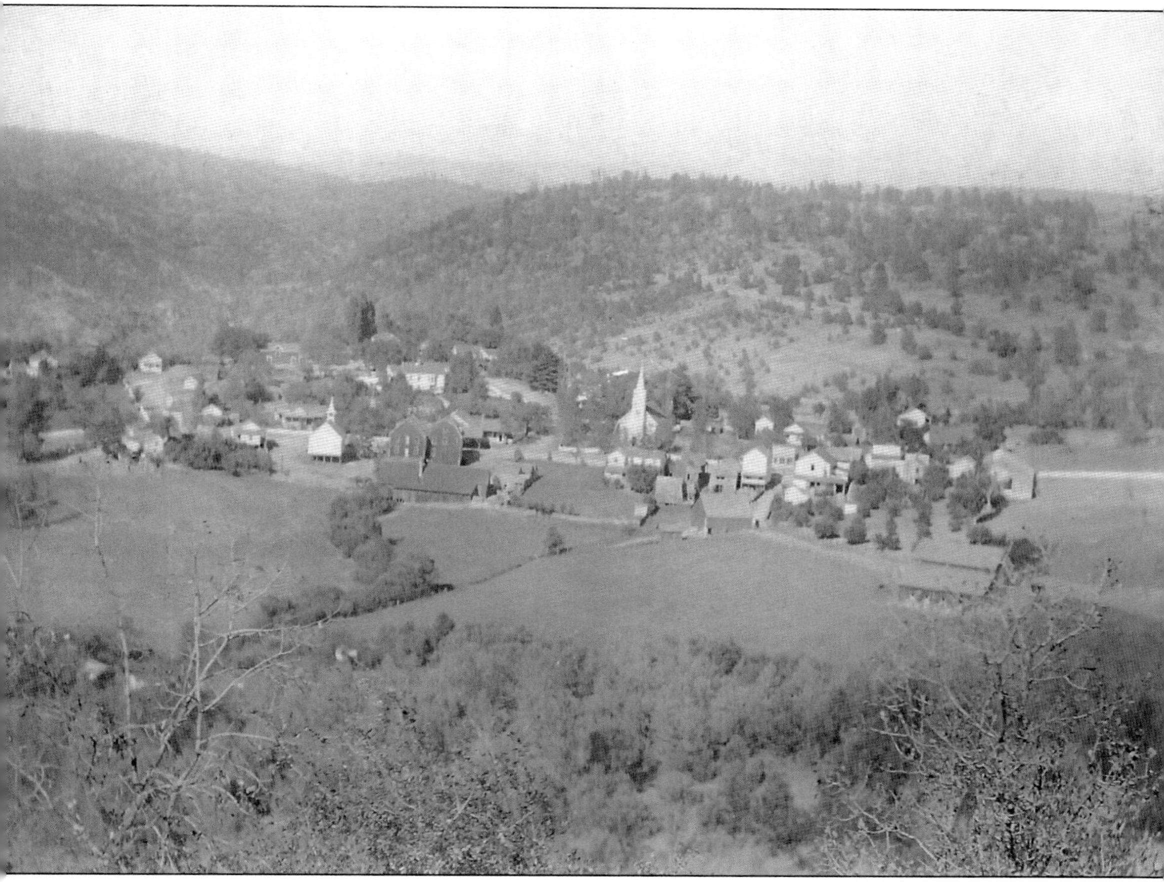

services, and many homes, moved to Smartsville. This panorama shows the town from one end to the other, beginning with the schoolhouse on the west (left) end, through a residential and agricultural area, and continuing through the town's business section with two churches, several hotels, stores, saloons, and livery stable. (Author's collection.)

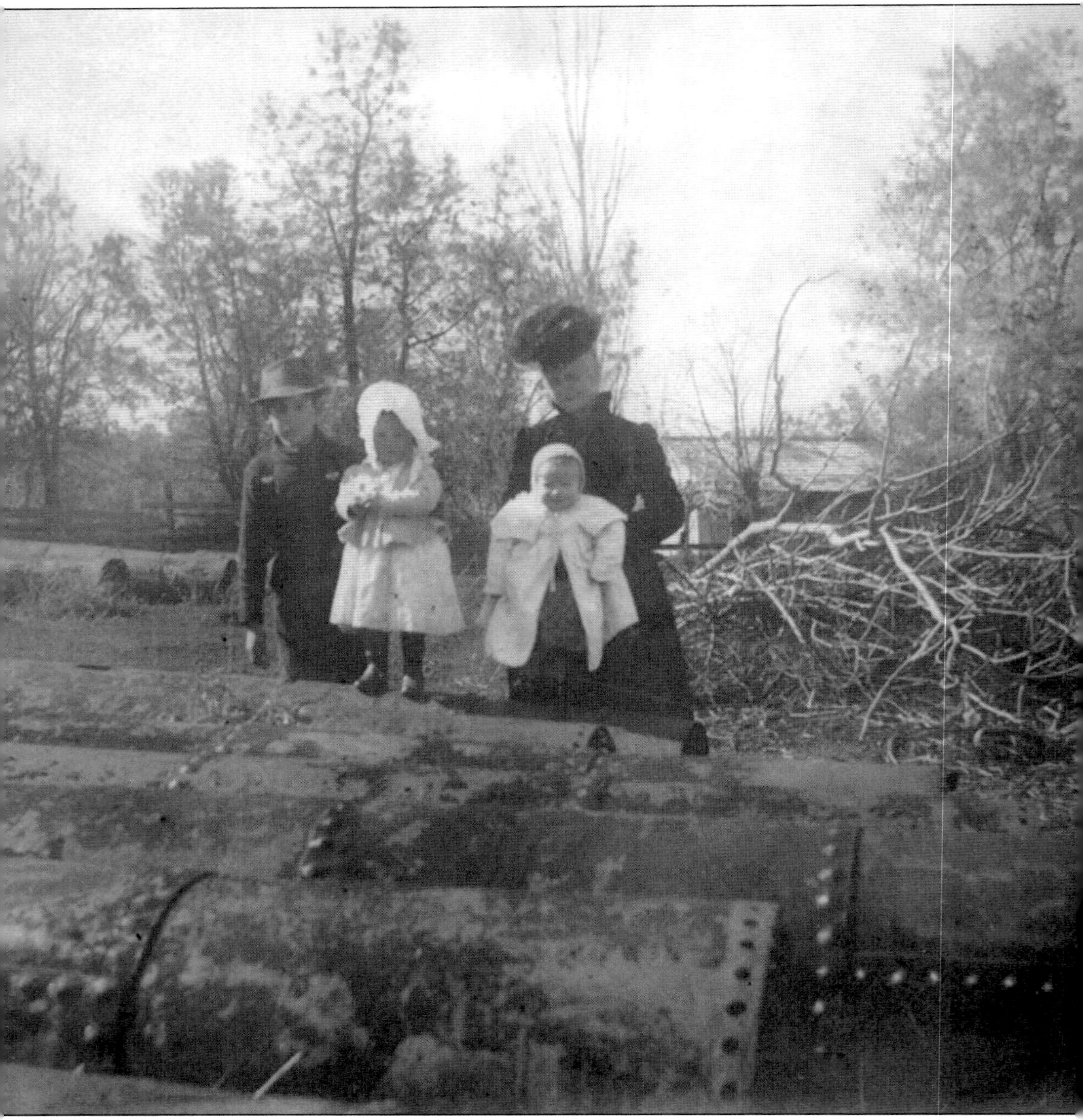

The pipes for the Excelsior Mine were manufactured in Smartsville. The pipes were made of sheet iron and were riveted together. These pipes are stacked on the side of the road, waiting to be delivered. Martha McConnell, Smartsville's druggist, posed for this photograph with her children standing on the pipes. (Courtesy of Mollie Plitzco.)

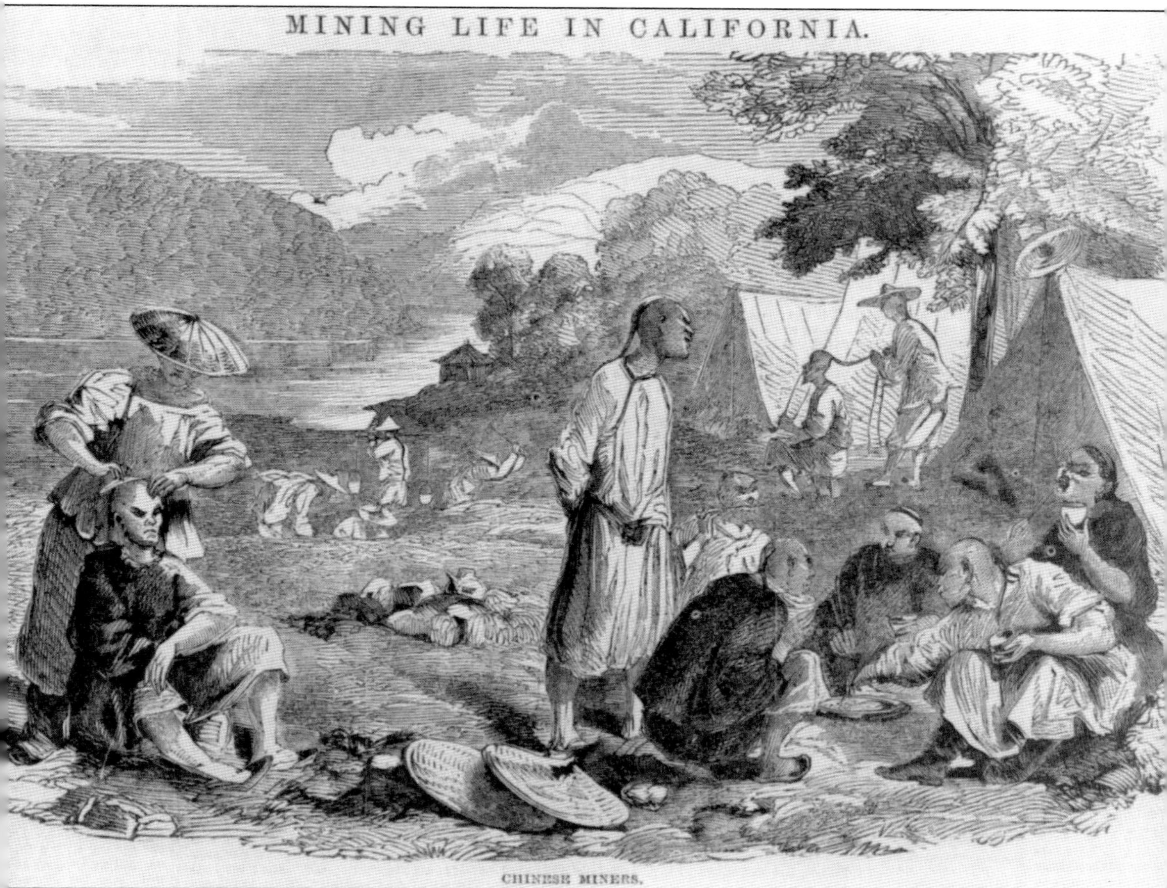

The Chinese were also part of the Yuba mining scene. They came seeking their fortunes, just like the other settlers, and their culture worked to their advantage. When building his ditches, James O'Brien found that he had to hire the whole group, or none of them would work for him. Because they could not own land, these immigrants leased already worked tailings and reworked them thoroughly in order to get the remaining gold. Besides their work ethic and frugality, they grew their own vegetables and ate a more healthy diet then the typical miner. Some also grew and sold fresh vegetables to the settlers. (Courtesy of Library of Congress.)

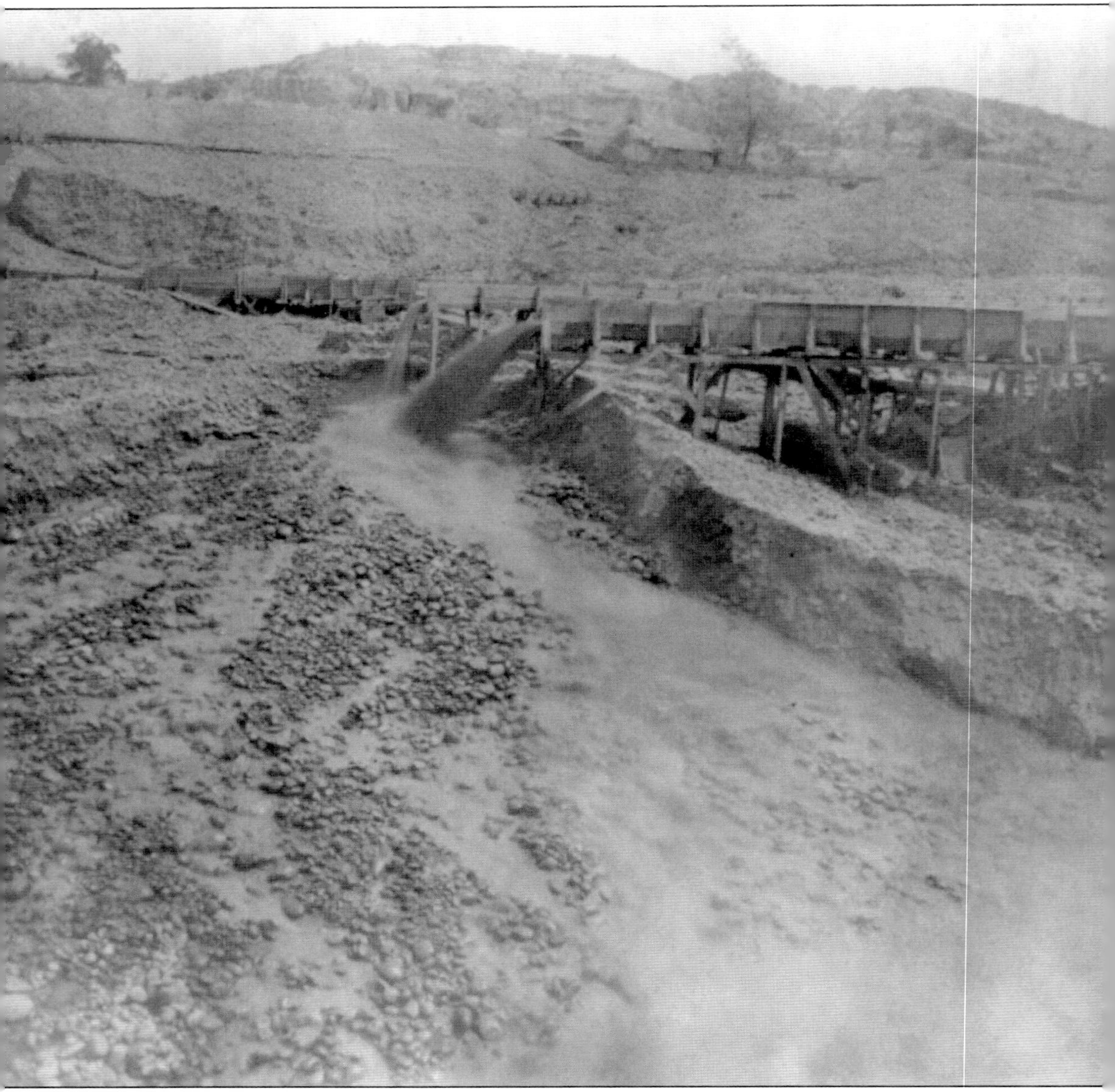

At the end of a hydraulic mining tunnel, tail flumes discharged the tailings into the river. The tail flumes were built in 12-foot sections, four feet wide and three feet deep. More sections could be added as needed. The flume also has riffles in it to catch gold. The tailings—also called the slurry, or slickens—eventually covered over previous town sites like Rose's Bar. (Courtesy of Library of Congress.)

Four

Mining vs. Agriculture

California's First Environmental Law

Hydraulic mining disposed of debris by dumping it in the Yuba River. When strong rains came, the sludge washed downstream, clogging the river. As early as 1856 there were concerns about the tailings, but in 1862, torrential rains came and the water and tailings overflowed the confines of the rivers, flooding Marysville and inundating the farmers' fields with silt, causing inconceivable damage. The storms of 1875 caused the flooding to overtop the levees in Marysville and filled the inside of the levees like a bowl. Marysville recovered with the contributions of many, including some mine owners. James O'Brien directed his crew to work around the clock building higher levees. However, the farmers' land was ruined by the slickens that covered it, and appeals on their behalf to the state legislature were unsuccessful.

The miners in Yuba were in a particularly bad spot. Tailings came from the hydraulic mines in Nevada County too, so the miners of Smartsville and Timbuctoo tried unsuccessfully to be annexed by Nevada County. Feeling threatened, the mine owners in both counties formed the Hydraulic Miners Association in 1876 to defend against any suits that would surely be forthcoming. It became clear to the farmers that this was not going to be an easy or an inexpensive battle, and so on August 24, 1878, the Anti-Debris Association was formed. Its mission was to prosecute cases that would challenge the rights of the miners to use the rivers for dumping tailings. One such case, Edwards Woodruff v North Bloomfield Gravel Mining Company, was filed in 1882. On January 7, 1884, after two years of litigation in the case and over 2,000 witnesses and 20,000 pages of written testimony, Judge Lorenzo Sawyer handed down an opinion now known as the Sawyer Decision. It banned any mining process that allowed tailings to be dumped into the tributaries of the Sacramento River. The culmination of this case is considered to be a landmark legal decision touted as California's first environmental law.

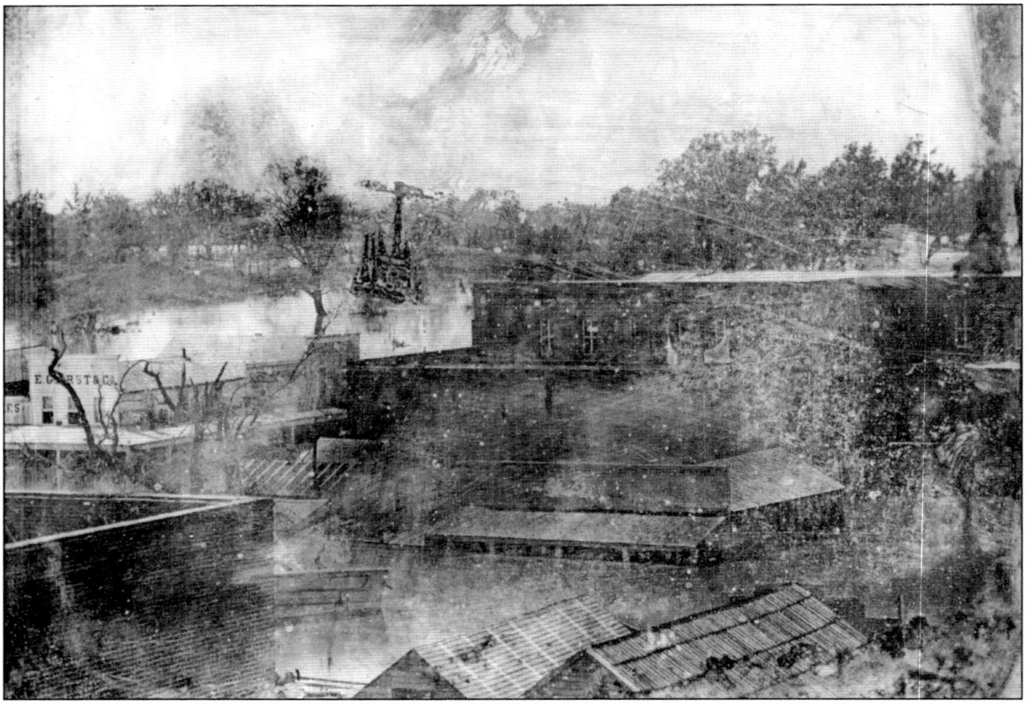

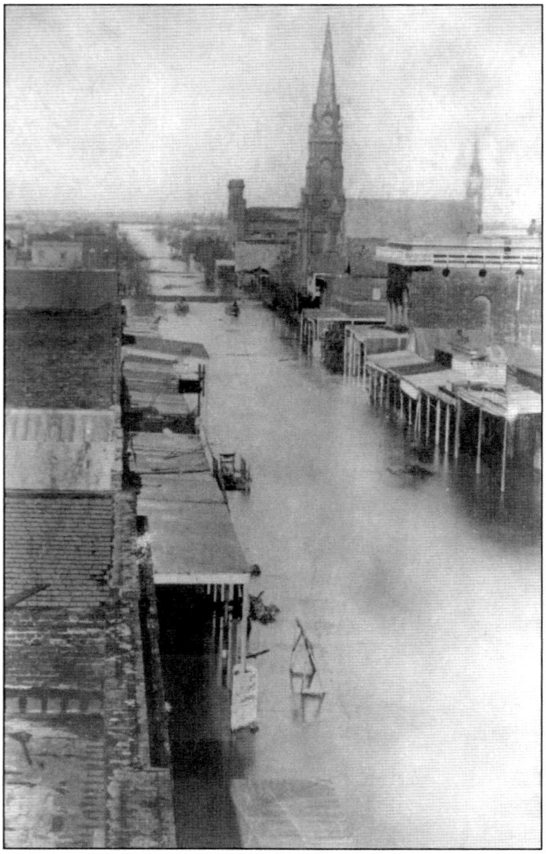

It is likely that this 1853 flood was caused by more than just the high amounts of rainfall. The townspeople may have not yet known that the silt washing through rivers was making it more difficult for the rainfall to drain, as we now understand. They did, however, know that the bars of the Yuba were already beginning to become covered in debris.

By 1867, the riverbeds were as high as the streets of Marysville, leading to the construction of several levees. Complaints from local farmers were given little consideration, as much of Yuba and Sutter Counties depended on the hydraulic mines economically. This photograph of the 1867 Marysville flood shows that the water rose well over the boardwalks and entered the businesses.

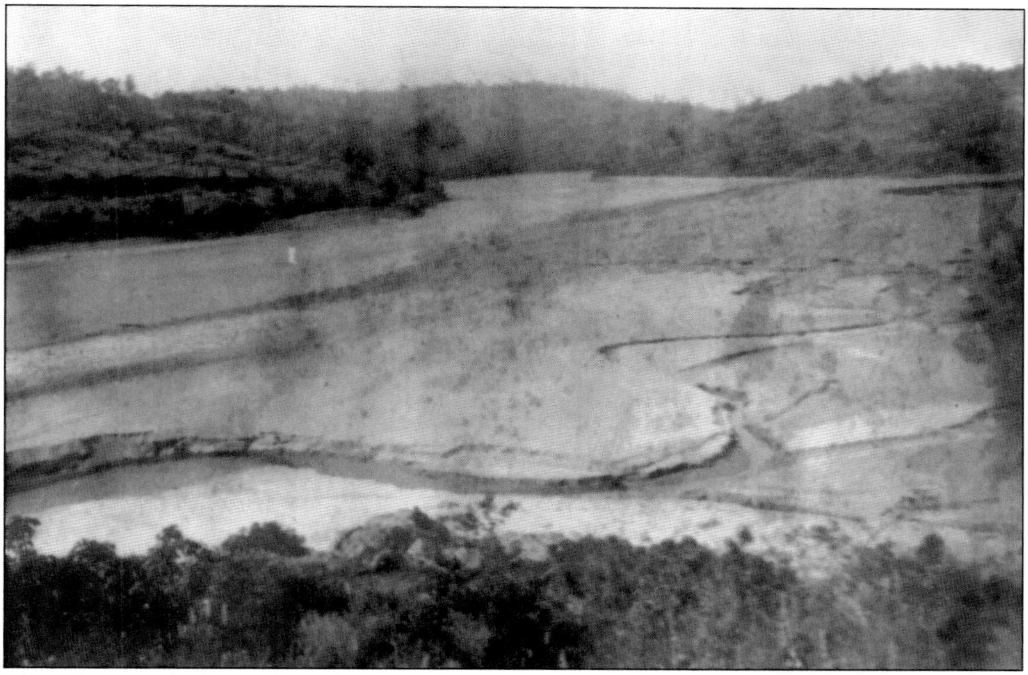

This photograph shows the tailings piled high around the Yuba River. It is easy to see how the rains could wash them downstream along with the rushing storm water, and this is just one spot on the river. There were more than seven tunnels running into the river within a few miles of this bank—all of them adding to the quantity of tailings. The mines farther upriver in Nevada County contributed as well. (Courtesy of Library of Congress.)

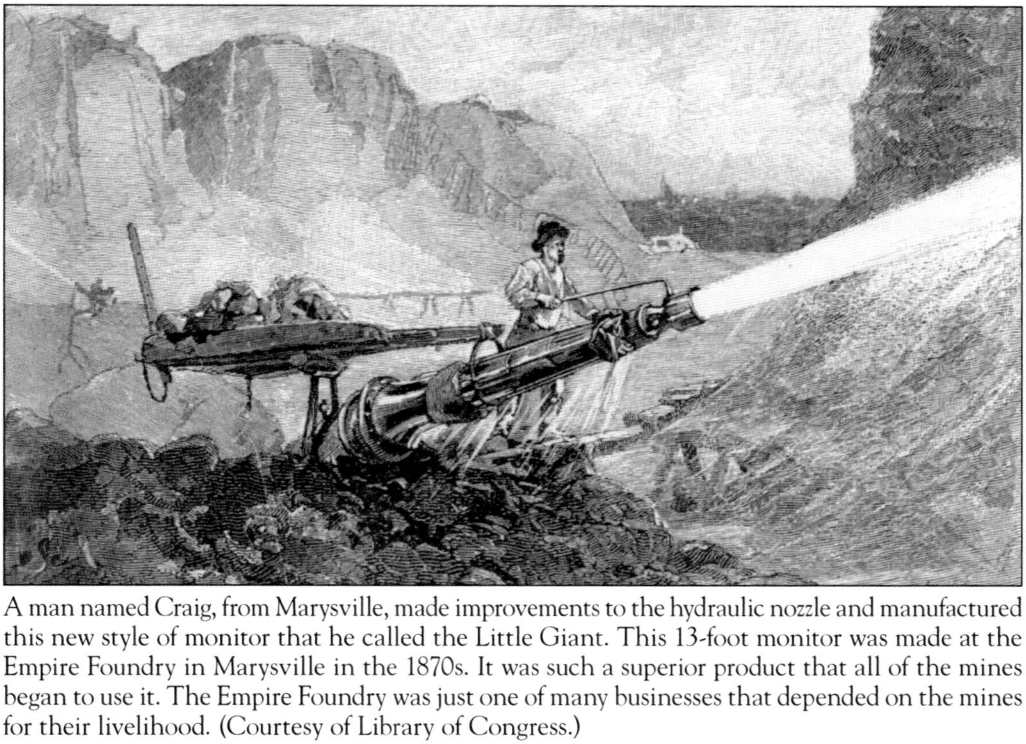

A man named Craig, from Marysville, made improvements to the hydraulic nozzle and manufactured this new style of monitor that he called the Little Giant. This 13-foot monitor was made at the Empire Foundry in Marysville in the 1870s. It was such a superior product that all of the mines began to use it. The Empire Foundry was just one of many businesses that depended on the mines for their livelihood. (Courtesy of Library of Congress.)

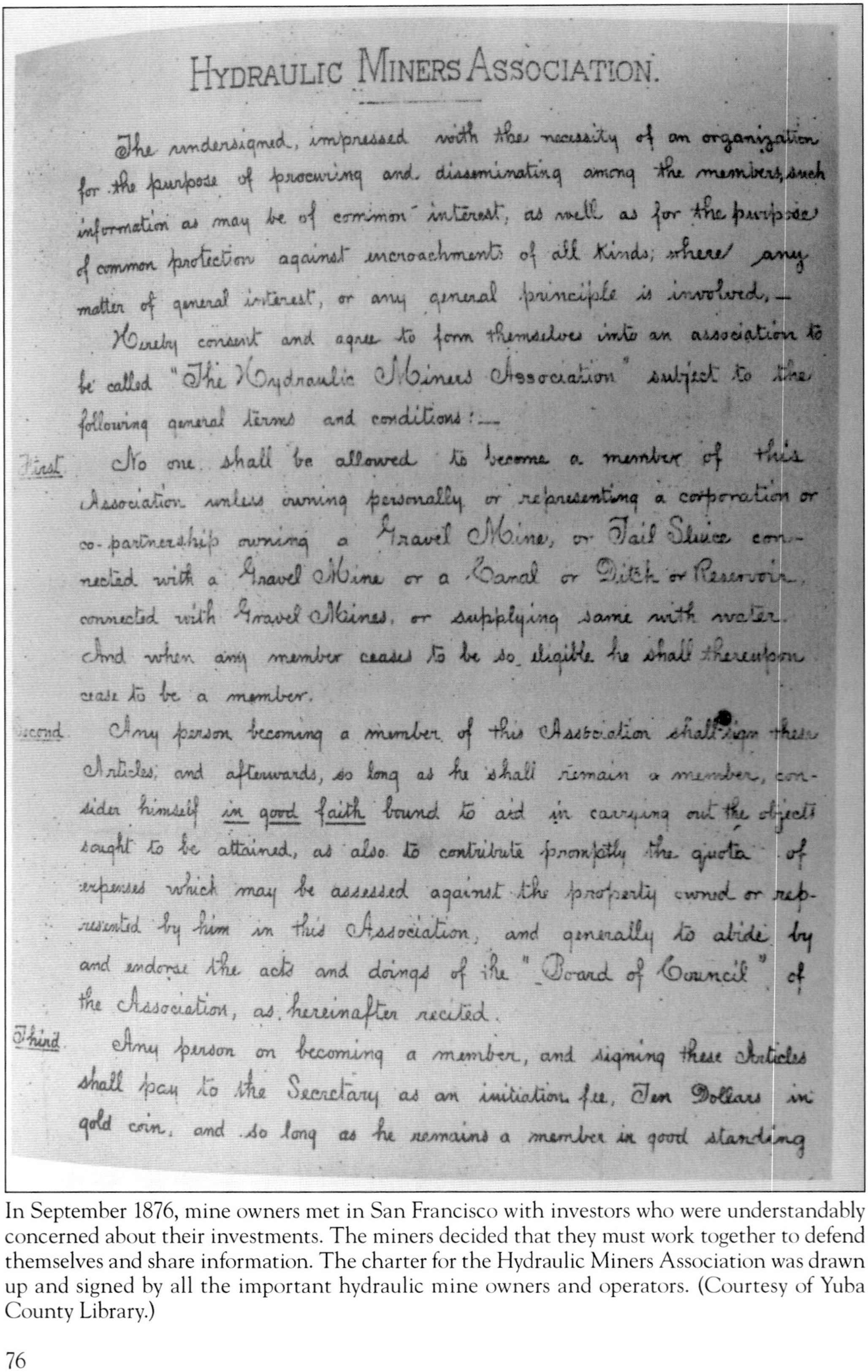

HYDRAULIC MINERS ASSOCIATION.

The undersigned, impressed with the necessity of an organization for the purpose of procuring and disseminating among the members such information as may be of common interest, as well as for the purpose of common protection against encroachments of all kinds; where any matter of general interest, or any general principle is involved,—

Hereby consent and agree to form themselves into an association to be called "The Hydraulic Miners Association" subject to the following general terms and conditions:—

First. No one shall be allowed to become a member of this Association unless owning personally or representing a corporation or co-partnership owning a Gravel Mine, or Tail Sluice connected with a Gravel Mine or a Canal or Ditch or Reservoir, connected with Gravel Mines, or supplying same with water. And when any member ceases to be so eligible he shall thereupon cease to be a member.

Second. Any person becoming a member of this Association shall sign these Articles; and afterwards, so long as he shall remain a member, consider himself in good faith bound to aid in carrying out the objects sought to be attained, as also to contribute promptly the quota of expenses which may be assessed against the property owned or represented by him in this Association, and generally to abide by and endorse the acts and doings of the "Board of Council" of the Association, as hereinafter recited.

Third. Any person on becoming a member, and signing these Articles shall pay to the Secretary as an initiation fee, Ten Dollars in gold coin, and so long as he remains a member in good standing

In September 1876, mine owners met in San Francisco with investors who were understandably concerned about their investments. The miners decided that they must work together to defend themselves and share information. The charter for the Hydraulic Miners Association was drawn up and signed by all the important hydraulic mine owners and operators. (Courtesy of Yuba County Library.)

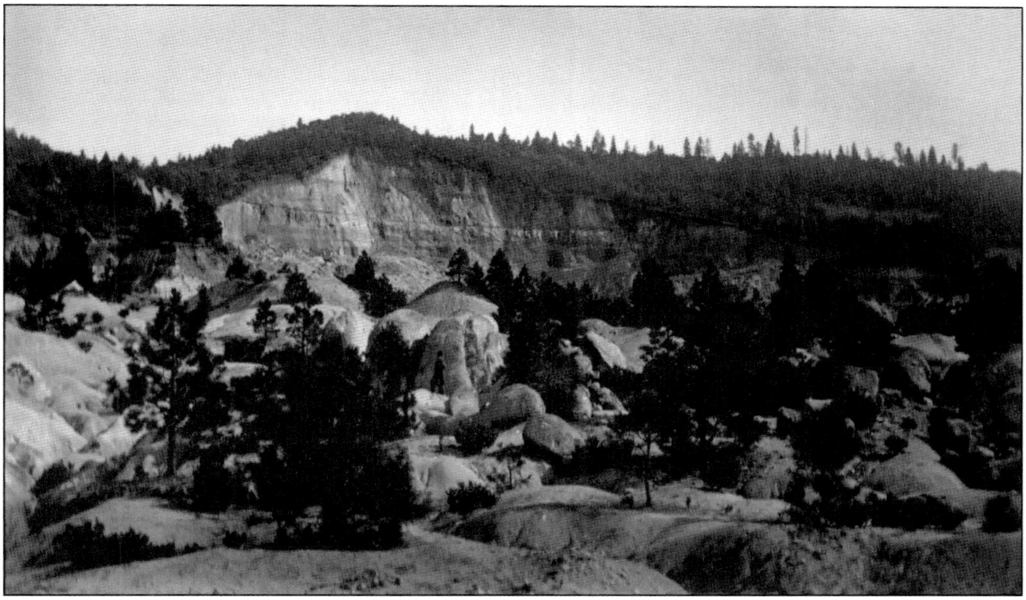

Both of these photographs show land involved a suit against the Miners Association and that served as exhibits one and two in the defense. They show the destruction and scars caused by hydraulic mining, but not the real devastation caused by the disposal of the materials that were washed away. (Both, courtesy of National Archives and Records Administration.)

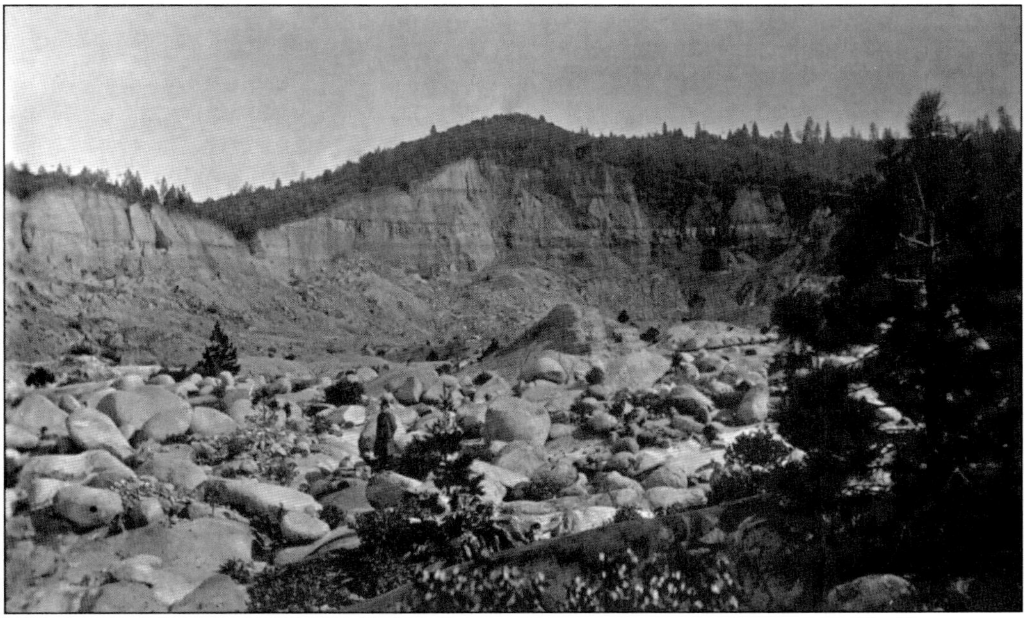

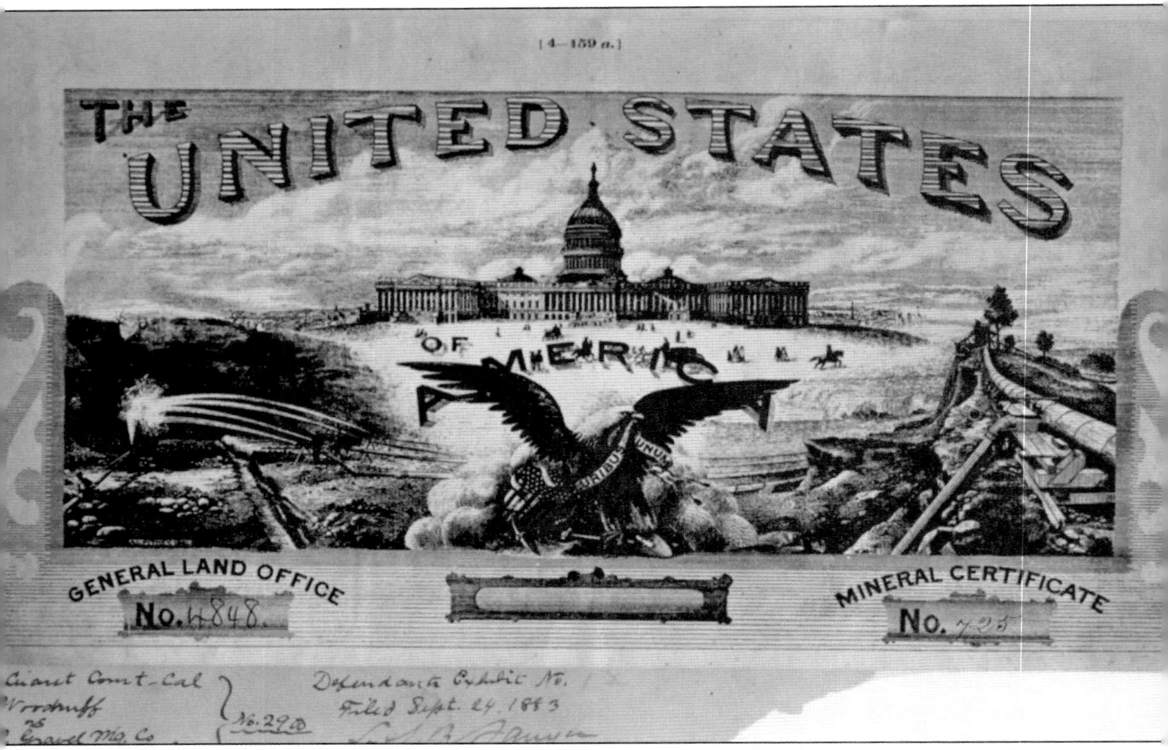

This mineral certificate was used as an exhibit in the Woodruff case to show that the miners did have mineral rights as granted by the federal government. They also held federal land patents as well. Up until this time, no one had considered that what a property owner did with his land should not infringe on another's rights. This was also the rationale for asking that the case be heard before a federal court. The farmers used Edwards Woodruff's name, without his knowledge, as that of the plaintiff, so that they could petition the federal courts for an injunction against hydraulic mining. This was possible because, even though he owned property in Yuba, he was a citizen of New York. Woodruff must not have minded because, while visiting London, he was approached by someone offering him a sum of money many times the value of his property in California to withdraw the petition. He refused. (Courtesy of National Archives and Records Administration.)

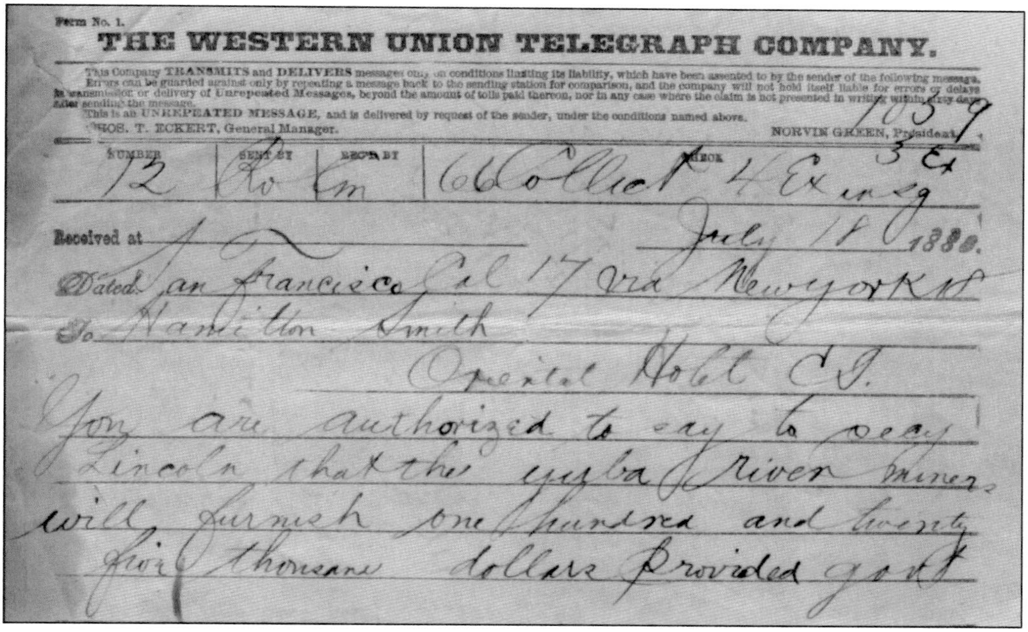
Form No. 1.

THE WESTERN UNION TELEGRAPH COMPANY.

This Company TRANSMITS and DELIVERS messages only on conditions limiting its liability, which have been assented to by the sender of the following message. Errors can be guarded against only by repeating a message back to the sending station for comparison, and the company will not hold itself liable for errors or delays in transmission or delivery of Unrepeated Messages, beyond the amount of tolls paid thereon, nor in any case where the claim is not presented in writing within sixty days after sending the message.
This is an UNREPEATED MESSAGE, and is delivered by request of the sender, under the conditions named above.
THOS. T. ECKERT, General Manager. NORVIN GREEN, President. 1059

NUMBER	SENT BY	REC'D BY	CHECK
12	Ro	m	66 Collect 4 Exng 3 Ex

Received at ___ July 18 1880.

Dated San Francisco Cal 17 via New York 18

To Hamilton Smith

Oriental Hotel C.I.

You are authorized to say to Secy Lincoln that the Yuba river miners will furnish one hundred and twenty five thousand dollars provided govt

These telegrams from July 1880 pledge that the miners would offer $125,000 to build a debris dam, if the government would pledge the same. In 1880, the State of California constructed a dam across the Yuba River—about two miles in length and located about eight miles east of Marysville—at a cost of over $200,000. This dam failed the following winter. The state and federal governments then appropriated $800,000 for dams on the Yuba River, with federal engineers expending the money. They first constructed a well-built dam of logs and stone, but that dam failed the next winter as well. They then built two concrete dams across the river, but they were completely destroyed within a few years. In all, over $2 million was expended building dams on that river. (Courtesy of National Archives and Record Administration.)

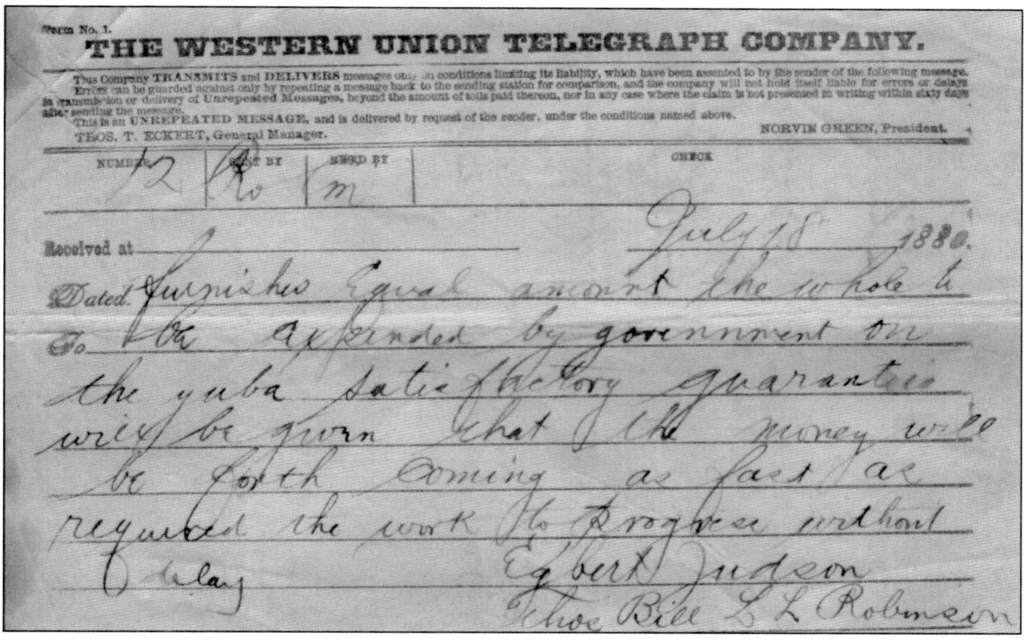
Form No. 1.

THE WESTERN UNION TELEGRAPH COMPANY.

This Company TRANSMITS and DELIVERS messages only on conditions limiting its liability, which have been assented to by the sender of the following message. Errors can be guarded against only by repeating a message back to the sending station for comparison, and the company will not hold itself liable for errors or delays in transmission or delivery of Unrepeated Messages, beyond the amount of tolls paid thereon, nor in any case where the claim is not presented in writing within sixty days after sending the message.
This is an UNREPEATED MESSAGE, and is delivered by request of the sender, under the conditions named above.
THOS. T. ECKERT, General Manager. NORVIN GREEN, President.

NUMBER	SENT BY	REC'D BY	CHECK
12	Ro	m	

Received at ___ July 18 1880.

Dated furnishes equal amount the whole to

To be expended by government on the Yuba satisfactory guarantee will be given that the money will be forthcoming as fast as required the work to progress without delay

Egbert Judson

Thos Bell L L Robinson

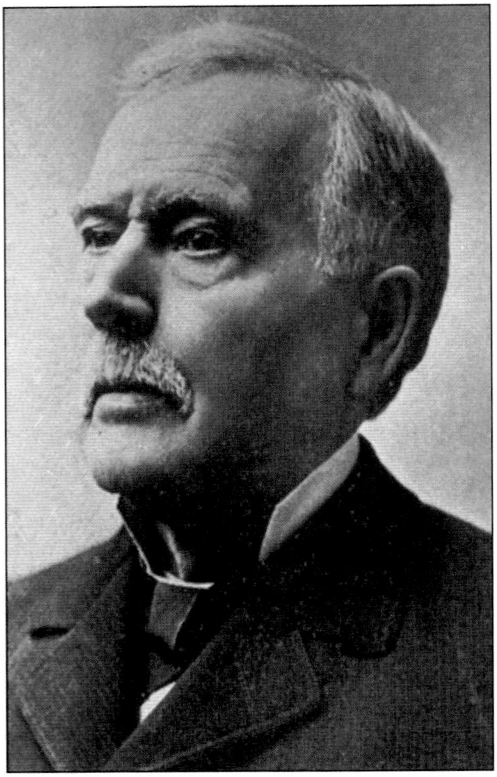

Two brothers, Isaac (left) and William Belcher (below), represented the opposing sides of the case, making things more confusing. Isaac Belcher, besides being a practicing attorney, had been district attorney in 1855 and a district judge in 1863. He also served a term as a justice on the California Supreme Court from 1872 to 1874, as vice-president of the California Constitutional Convention in 1879, and as commissioner of the Supreme Court from 1885 until his death in San Francisco in 1898. He was also a California State Library trustee and a trustee of Stanford University. William C. Belcher left Marysville and entered legal practice in San Francisco in 1875. He was grand master of Masons in California from 1862 to 1865 and was elected grand commander of Knights Templar in 1864. He died in San Francisco in 1895. (Both, courtesy of Tammy L. Hopkins)

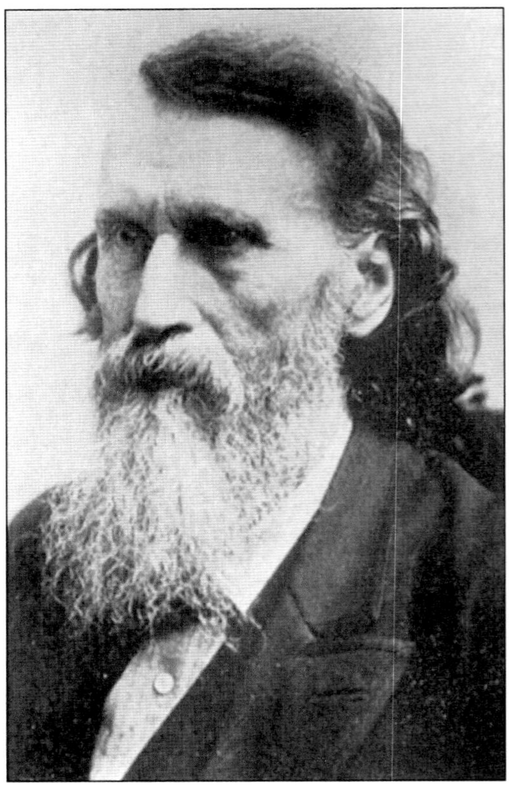

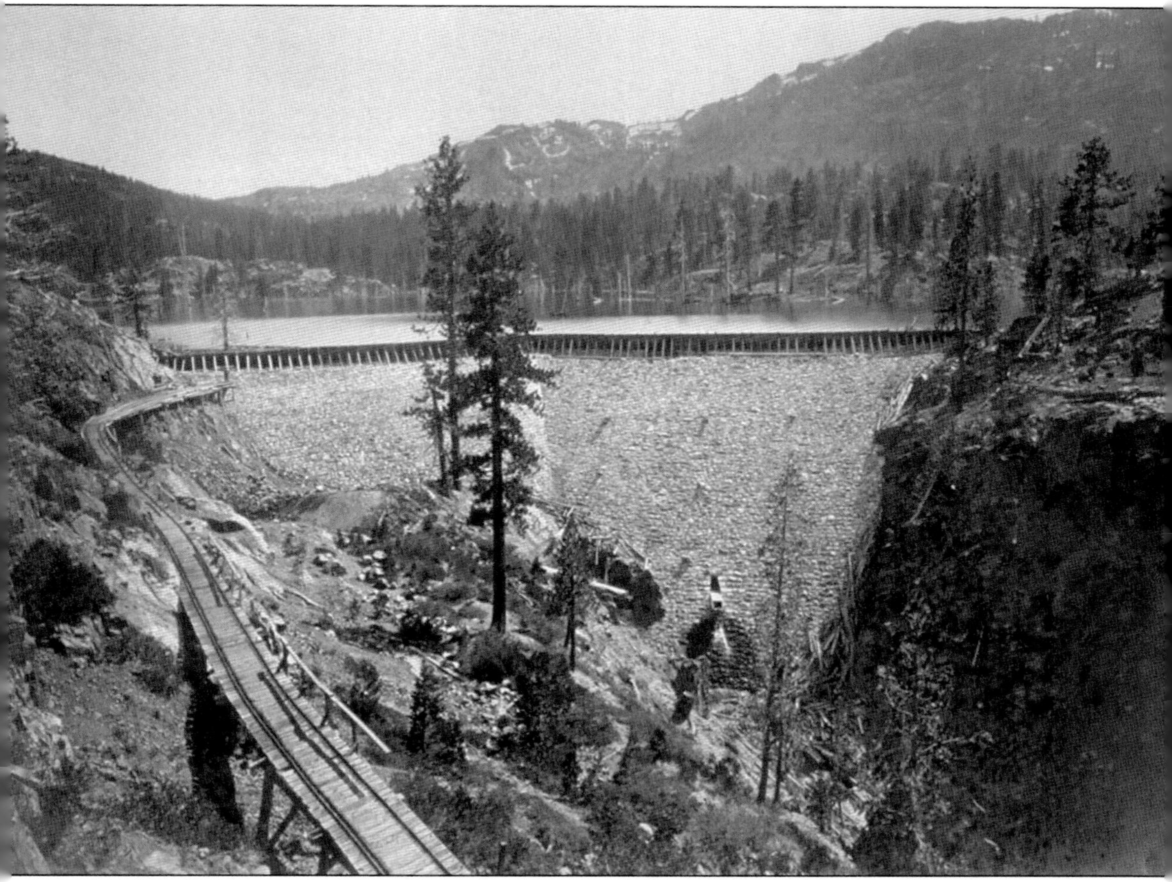

The English Dam was built from one bluff to another on the Yuba. Its purpose was to impound all of the tailings released above that point. One of the earliest civil engineering feats on the West Coast, the dam consisted of a 114-foot-tall reservoir constructed from a network of interlocking timber boxes filled with stones. As water filled the area enclosed by the dam, the trees were flooded—leaving white skeletal trunks projecting from the newly formed lake. (Courtesy of the Getty's Open Content Program.)

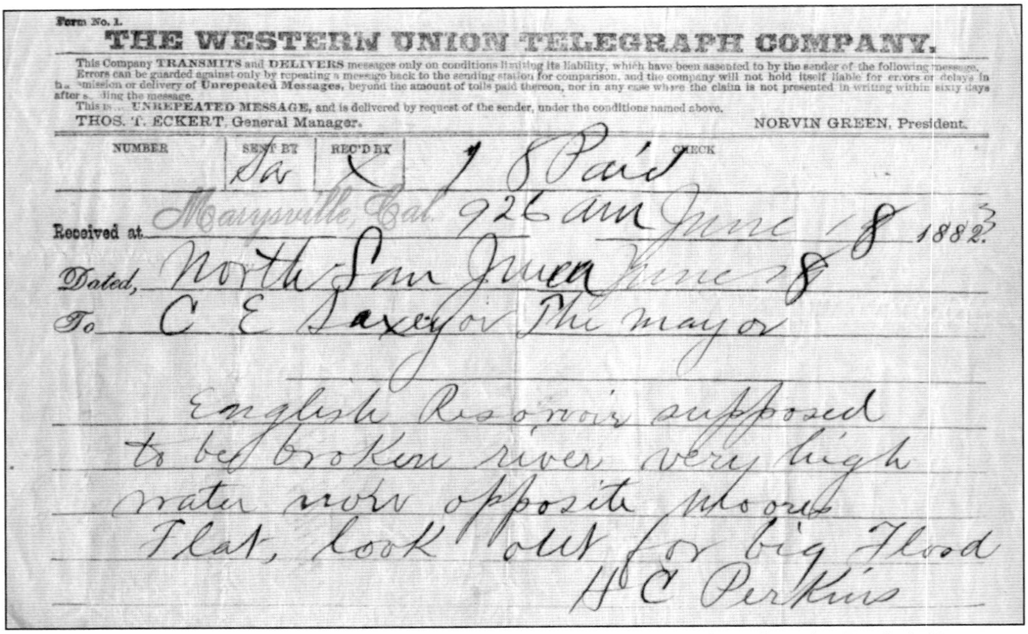

Form No. 1.

THE WESTERN UNION TELEGRAPH COMPANY.

This Company TRANSMITS and DELIVERS messages only on conditions limiting its liability, which have been assented to by the sender of the following message. Errors can be guarded against only by repeating a message back to the sending station for comparison, and the company will not hold itself liable for errors or delays in transmission or delivery of Unrepeated Messages, beyond the amount of tolls paid thereon, nor in any case where the claim is not presented in writing within sixty days after sending the message.
This is an UNREPEATED MESSAGE, and is delivered by request of the sender, under the conditions named above.

THOS. T. ECKERT, General Manager. NORVIN GREEN, President.

NUMBER	SENT BY	REC'D BY		CHECK
	Sa	X	1 8 Paid	

Received at Marysville, Cal. 926 am June 18 1883

Dated, North San Juan June 18

To C E Saxe or The mayor

English Reservoir supposed to be broken river very high water now opposite Moores Flat, look out for big Flood

H C Perkins

This telegram from June 18, 1883, warned residents of the valley below that the English Dam was breaking. It was suspected that the anti-debris faction had dynamited the dam in order to impel the courts to hasten a decision. It may have worked, too: less than six months later, a decision was handed down. (Courtesy of National Archives and Record Administration.)

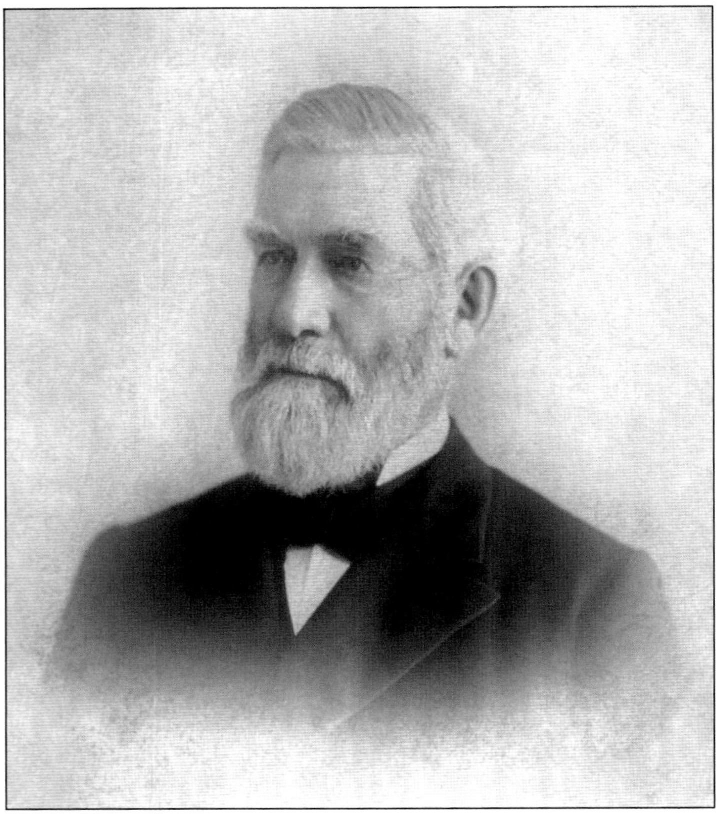

Judge Lorenzo Sawyer was responsible for the order known as the Sawyer Decision. It drastically curtailed hydraulic mining and imposed strict laws regarding any debris sent downstream. He first came to California in 1850 to mine but shortly fell back into practicing law. He first opened an office in Sacramento before going on to practice in Nevada City, and it was only later that he became a judge and heard this case. (Courtesy of Library of Congress.)

Five

ECONOMIC IMPACT

THE CAMINETTI ACT

Once hydraulic mining was stopped, the economy of the mining towns took a big hit. Surprisingly, the state economy saw a downturn as well. It seems that no one expected that without all of the money the mines put into the economy, the prices that farmers received for their goods would drop drastically. Additionally, capitalists were now shying away from investing in hard rock mines, which were also required to show that they did not contribute to the environmental problem. The miners tried desperately to see a bill passed that would allow hydraulic mines to build restraining dams and continue mining as long as they could keep their tailings from entering the waterways. In light of the economic downturn, support for their cause was gaining momentum. Another faction wanted the rivers to be reclaimed and rehabilitated, but the Sierra canyons were still full of debris, and each year rains washed more and more down into the tributaries, causing regular flooding. These groups eventually joined forces, and Congress established the Biggs Commission to investigate the debris problem. The commission findings recommended that miners be allowed to erect their own barriers, while determining that the government should build brush wing dams to scour the rivers. Anthony Caminetti introduced a bill that would act upon the recommendations of the Biggs Commission, and the Caminetti Act was signed into law in March 1893. This law gave the California Debris Commission the authority to regulate the resumption of hydraulic mining—under the provisions of the law—and made plans to improve the navigability of the rivers. The improvements to the Sacramento, Yuba, and Feather Rivers were made, allowing steamers to go as far north as Red Bluff and reach many river stops that had been inaccessible for over 20 years. Marysville was protected from future flooding by the Yuba cut-off, investors resumed investing, and some hydraulic mining resumed. Despite all this, the industry never again reached its boom levels.

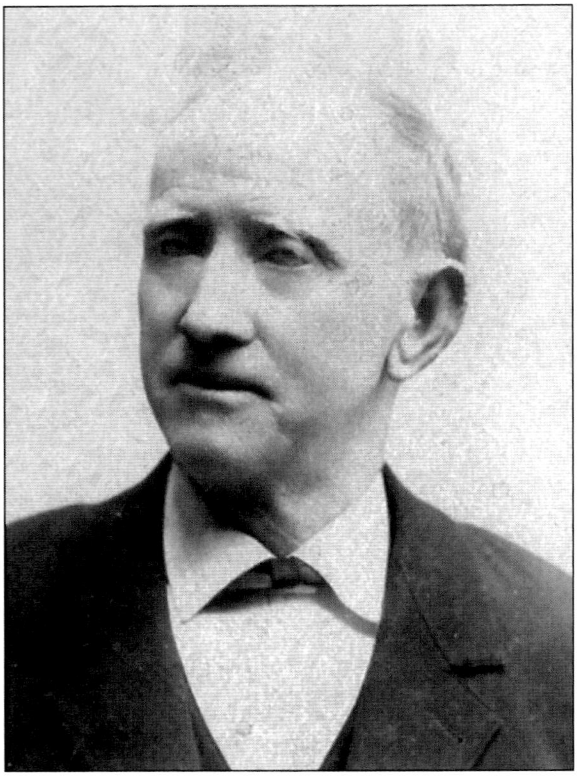

Patrick Campbell, owner of the Blue Point Mine, was infamous for not complying with the injunction against hydraulic mining. Even though he had been arrested and fined several times, he tried many ways of getting around the rules. One method was to entice the debris inspectors to stop at the saloon and have a drink, which prompted a lookout to report to the mine's foreman and have the water shut off. By the time the inspector arrived, no mining was being done. The article below, from a February 1885 *Daily Alta California* illustrates Campbell's infamy. He was so adept at using the law to keep from being held responsible for his actions that before this case had been resolved, the Yuba County Board of Supervisors had petitioned the governor for military intervention to bring him to justice. (Left, courtesy of Nevada County Historical Society; below, author's collection.)

A Relic of the Debris Case.

In November of last year Judge Keyser, of the Superior Court of Yuba county, inflicted a fine of $500, with the alternative of imprisonment, upon Patrick Campbell, then Superintendent of the Golden Gate Park Hydraulic Mining Company, for contempt of Court, committed by allowing debris from the mines of the company to flow into the Yuba river after an injunction had been issued prohibiting such action. Campbell refused to pay the fine, and being imprisoned, had sued out, through his attorneys, an application for a writ of habeas corpus, returnable before Judge Toohy. The matter was argued some time ago, Campbell's counsel claiming that Judge Keyser had not jurisdiction, and hence had exceeded his authority. Yeeterday Judge Toohy denied the petition, and remanded the prisoner.

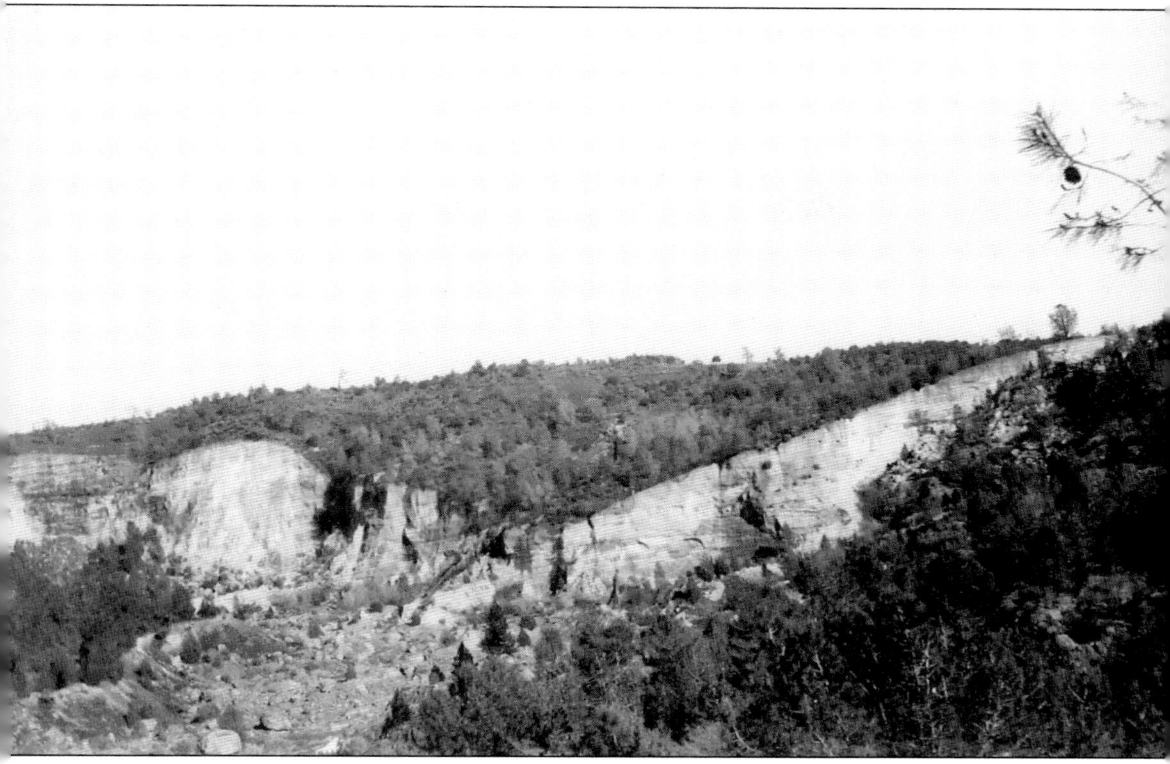

A malfunctioning dynamite blast at the Blue Point Mine caused this condition, known as the slide. When the power drift was blasted, it blew the targeted ground back into the hill rather than raising it upward. When water was applied and mixed with the clay behind the gravel deposit, it made the earth slide forward as the gravel was washed away in front of it. Patrick Campbell used this slide as a ruse to continue hydraulic mining—petitioning the Army Corp of Engineers to allow him to wash away enough gravel to uncover some equipment he claimed had been buried. They agreed, marking the area that he was allowed to wash with stakes. The stakes, of course, kept being moved to make it look as though they had not reached the point at which they were supposed to stop. (Courtesy of Helen McGovern.)

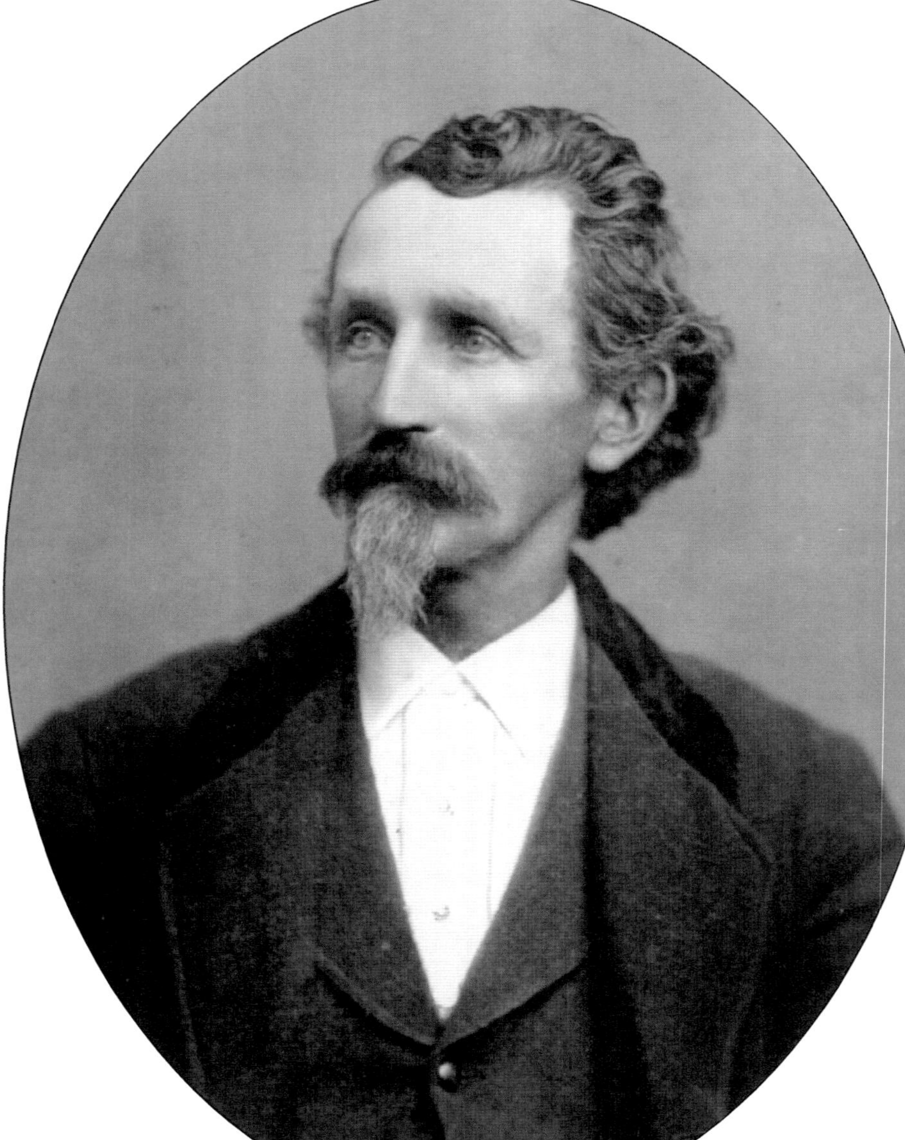

In 1887, Charles Compton (pictured) and W.W. Chamberlain reopened the Golden Gate Hydraulic Mine—also known as the Pat Campbell Mine—as a drift mine. In order to have drainage for the drift mine, the old hydraulic tunnel needed to be reopened by sluicing the residual material through the tunnel and into the ravine adjoining the river. They petitioned the Yuba County Board of Supervisors and consent was given. The new mine was supposed to employ at least 100 men year-round, revive things at Smartsville, create large trade for Marysville, and not damage the Yuba River more than any other quartz or drift mine. Hopes were dashed when the Marysville Drift Mining Company was forced to close down after only one year. The ratio of costs to returns from drift mining did not prove as profitable as those of hydraulic mining. The liabilities were $6,000—employees were owed $2,000, for which John Dunn, the manager, gave them a mortgage on his ranch. (Courtesy of Bud Compton.)

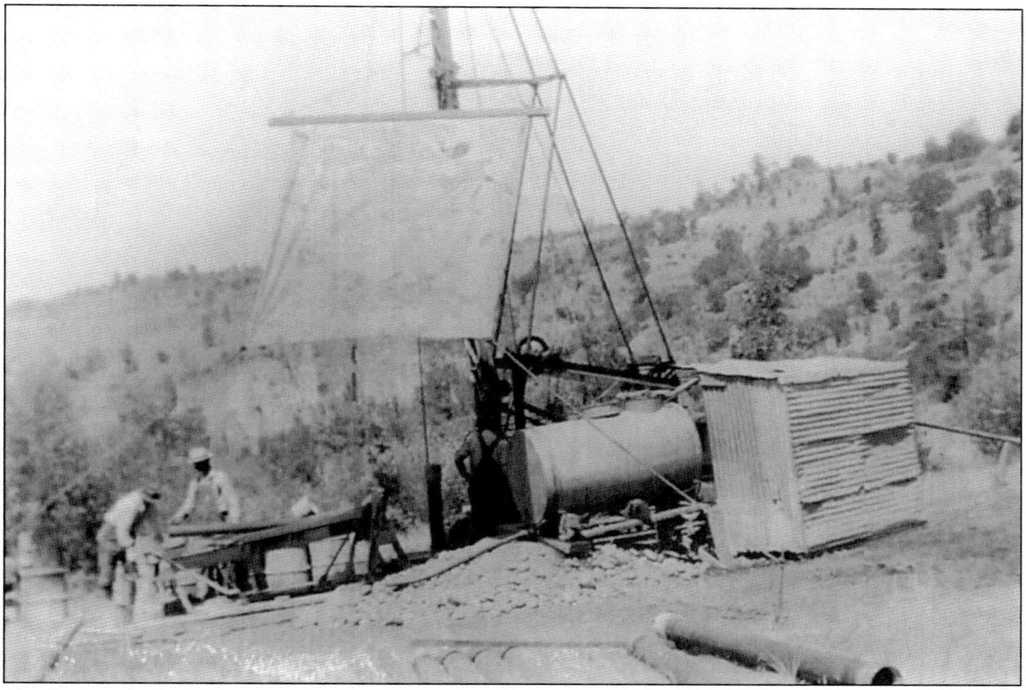

Working a drift mine takes as many as 10 men to every one in a hydraulic mine, and the progress is much slower. At this time, miners had some technology that could help in the process, such as the Keystone drill pictured above. Usually used to drill for petroleum, the miners used the drill to test the values of the drifts and decide where to dig. Similarly, the Pelton wheel donkey engine pictured below was considered a state-of-the-art power generator at the time. (Both, courtesy of Rosemary Freeland.)

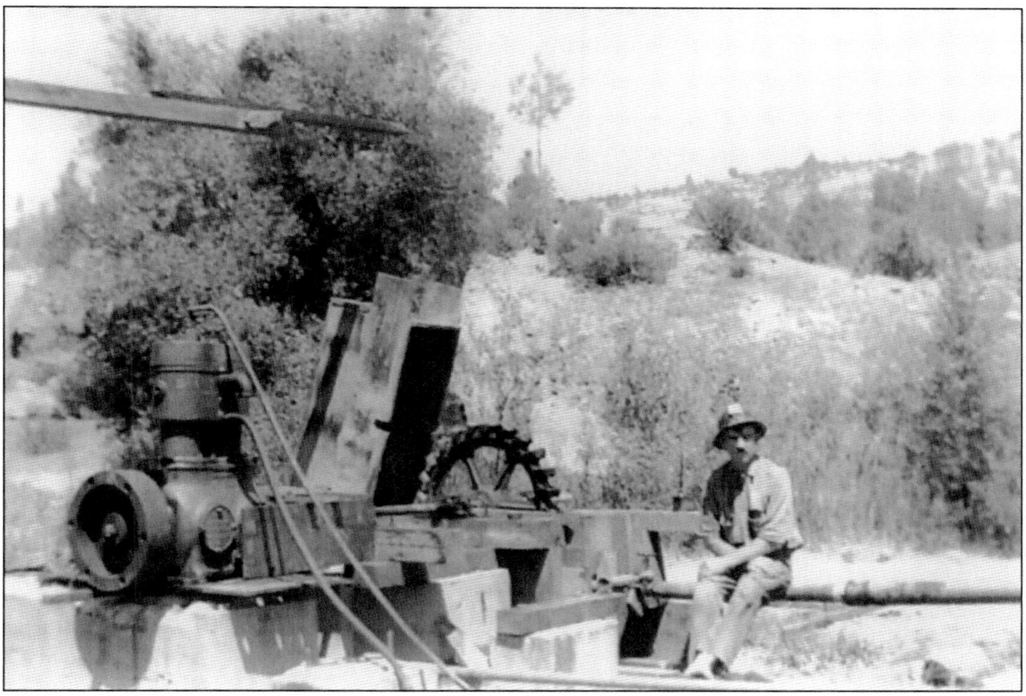

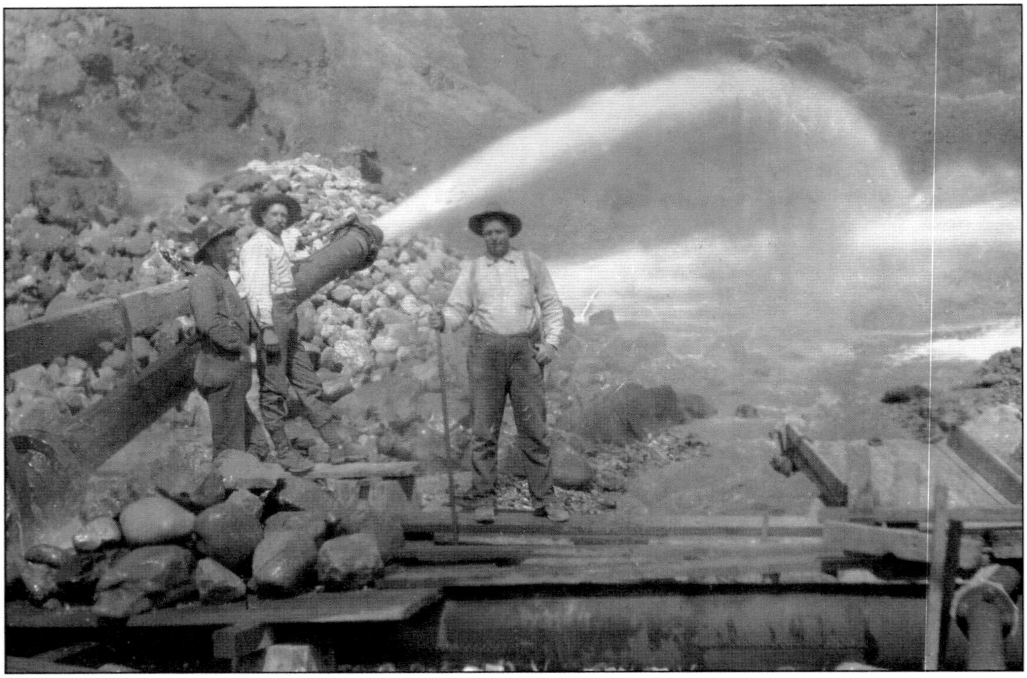

Hydraulic mining continued in some places, though whether it was done legally is not known. In this postcard from John A. Bean—sent to Mrs. P.C. Bean of Scales on November 17, 1911, and postmarked from Brandy City—Bean is at center, working the monitor, while Gene Covey holds the bar at right. (Courtesy of Yuba Feather Historical Society.)

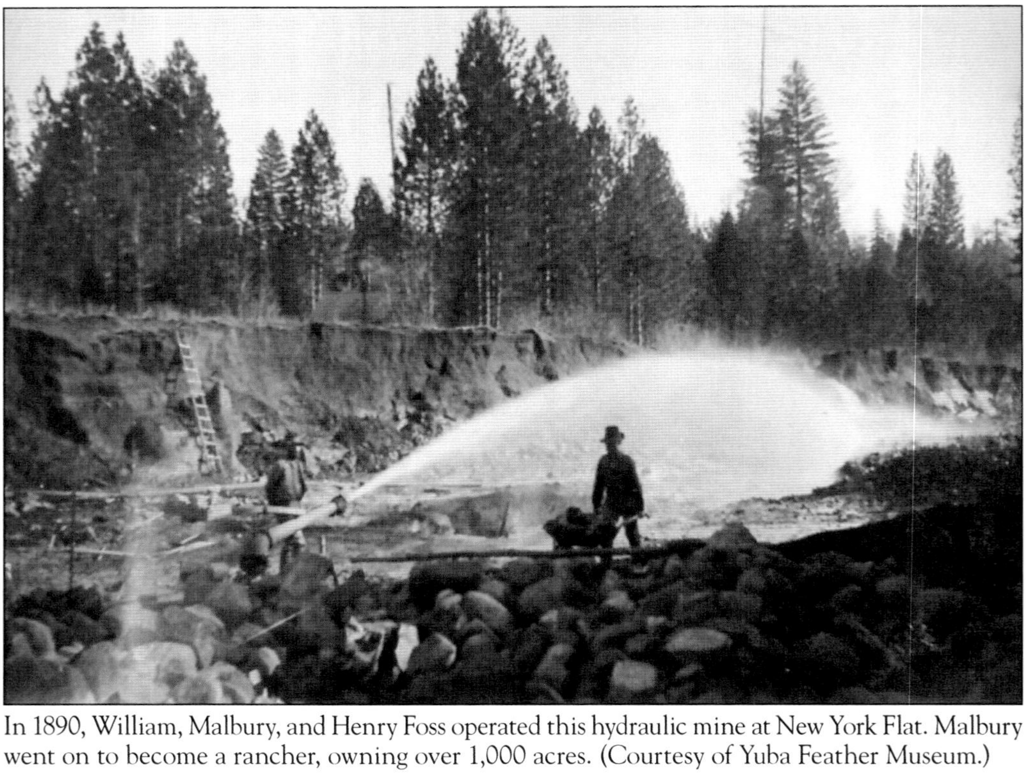

In 1890, William, Malbury, and Henry Foss operated this hydraulic mine at New York Flat. Malbury went on to become a rancher, owning over 1,000 acres. (Courtesy of Yuba Feather Museum.)

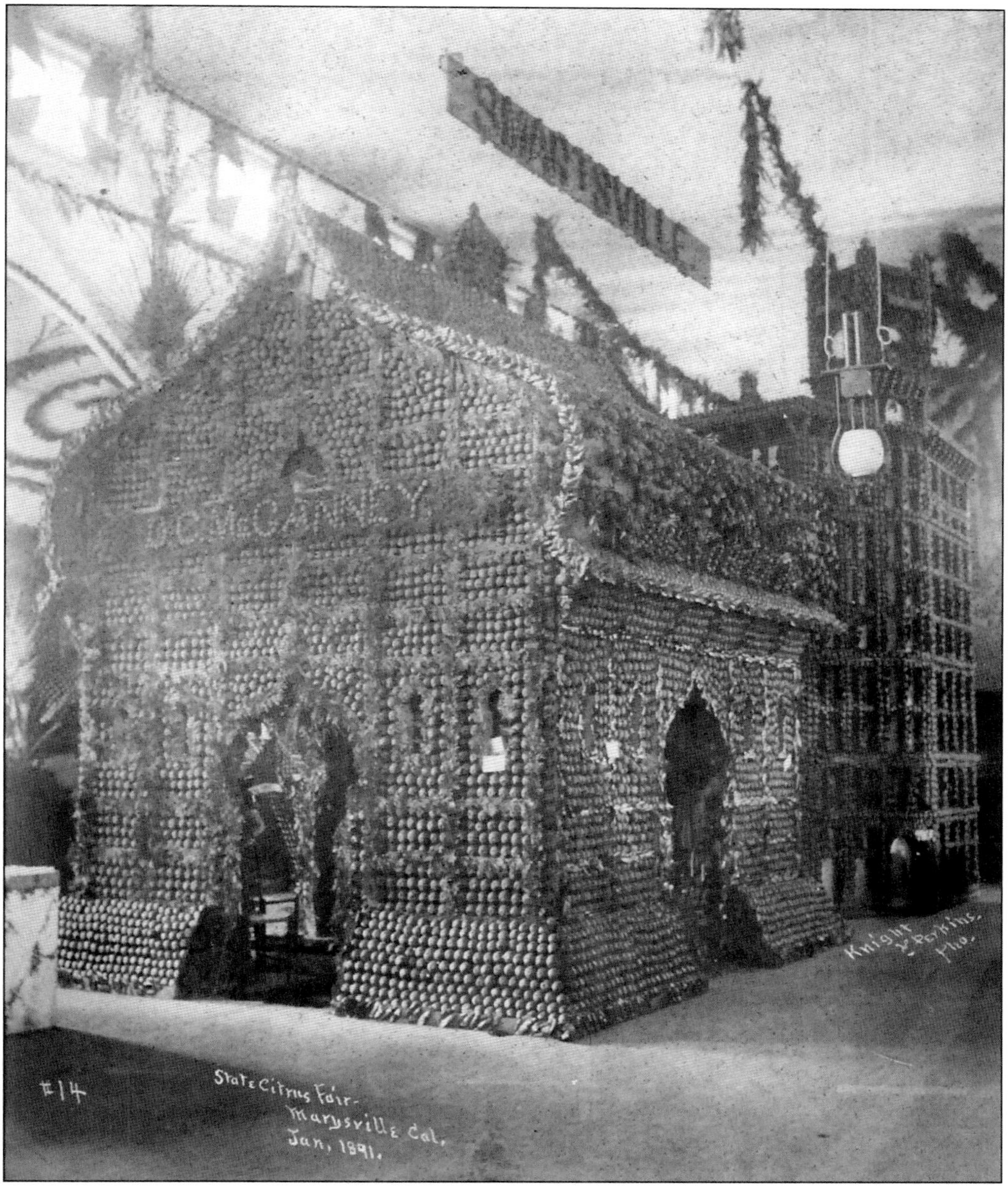

In 1886, an article titled "Irrigation Following Hydraulicking" appeared in the *Sacramento Union*. It stated that "In its vast reservoirs and hundreds of miles of ditches, hydraulic mining has left a legacy to the State that may, through the use of water for irrigation, accomplish much benefit. It is stated that the Excelsior Company at Smartsville is making arrangements for the employment of its copious flow of water in the irrigation of thousands of acres of foothill lands. Should orange culture be extensively engaged in upon the foothills, the water could be made to yield handsome revenue to the company." In January 1888, James O'Brien repurposed his ditches as agricultural waterways in the foothills. He planted and irrigated groves of citrus trees on his ranch, and other former mine operators soon did the same. This Moorish temple made entirely of locally grown citrus was on display at the State Citrus Fair, held in Marysville in January 1891. It features the produce of Smartsville growers and bears the names of two former mining men: Daniel McGanney and James O'Brien. (Courtesy of Yuba County Library.)

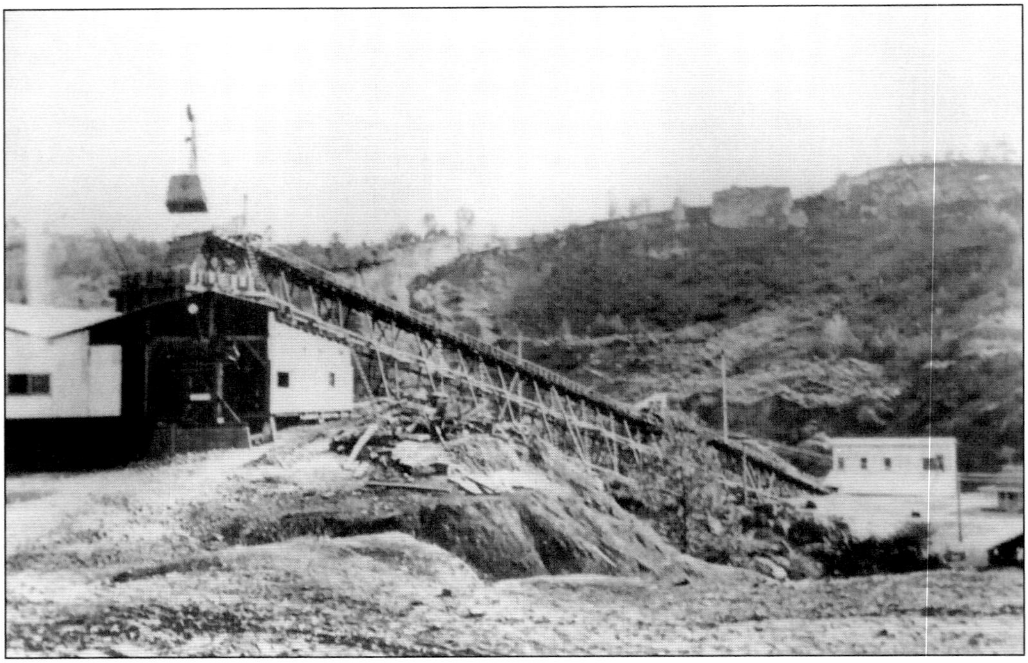

In 1911, under the direction of Ernest Tarr, another scheme at the Blue Point Mine attempted to use hydraulic mining again. Gravel was washed to the bucket line of a stationary dredger (seen above). After the gravel was washed and processed through the indoor concentrating plant (pictured below) it was transported by a conveyor belt up to a large ore bin on the rim of the channel. (Both, courtesy of Mary Clark.)

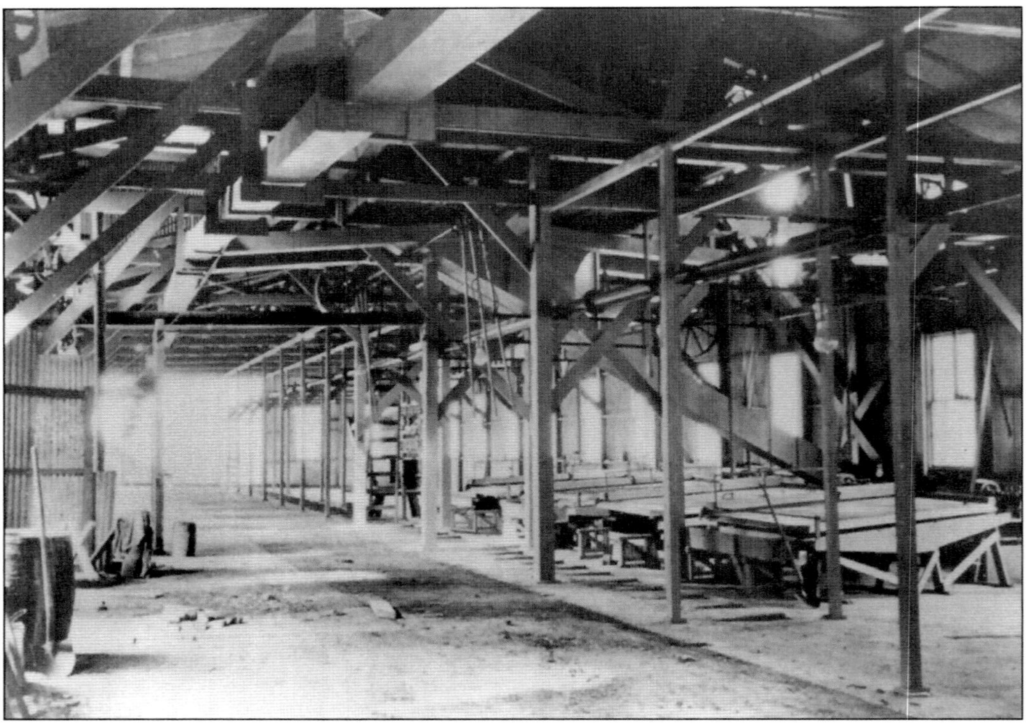

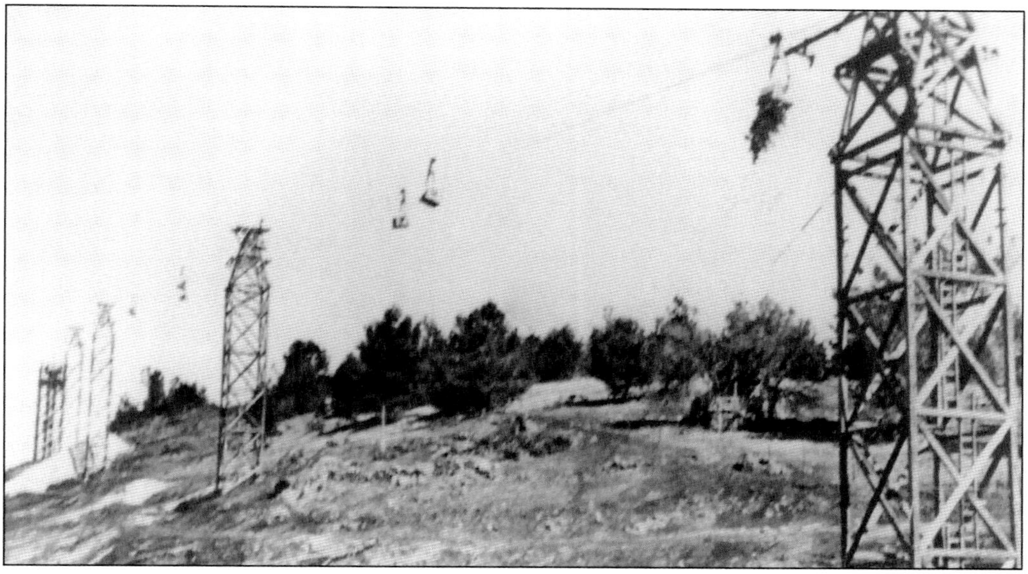

From the ore bin, the Tarr Mine's waste was loaded into one-yard buckets on an aerial tramway that consisted of 90-foot wooden towers. The buckets were made so they could be dumped anywhere along the line. Stock was sold to finance this very expensive operation, and many men were hired to build the exorbitantly expensive plant. After operating for a short time, the mine was closed, the corporation was dissolved, and the shareholders lost their investment. (Above, courtesy of Mary Clark; below, courtesy of Brian Bisnett.)

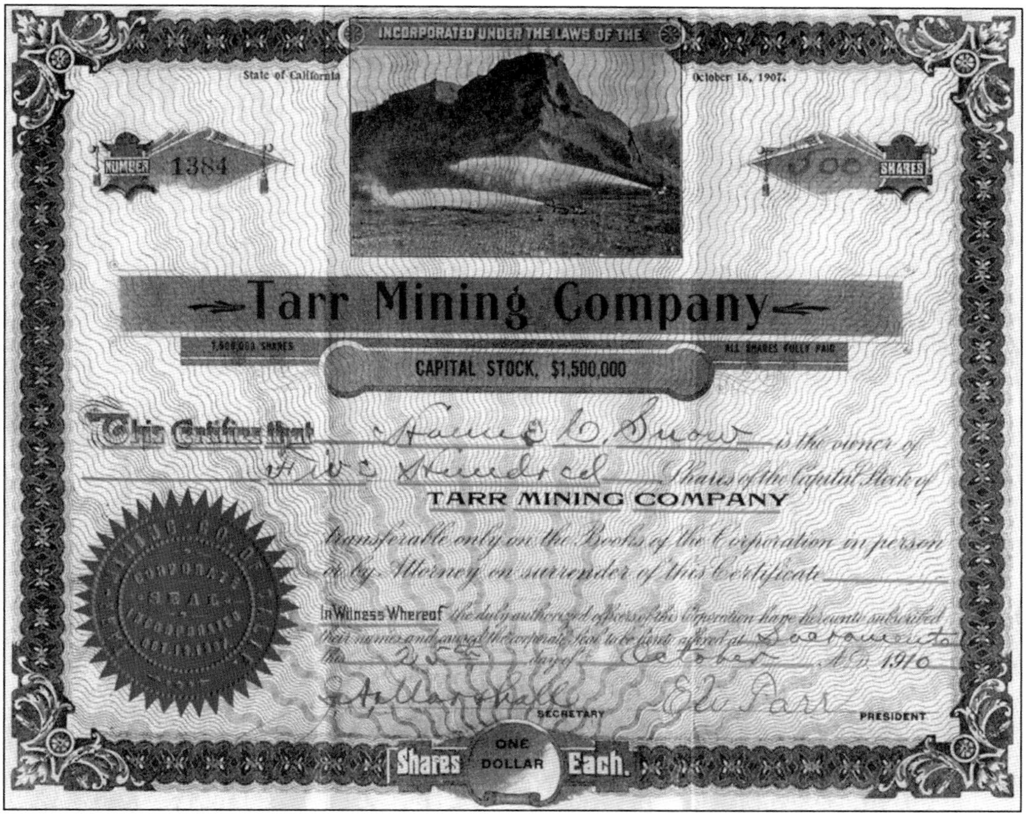

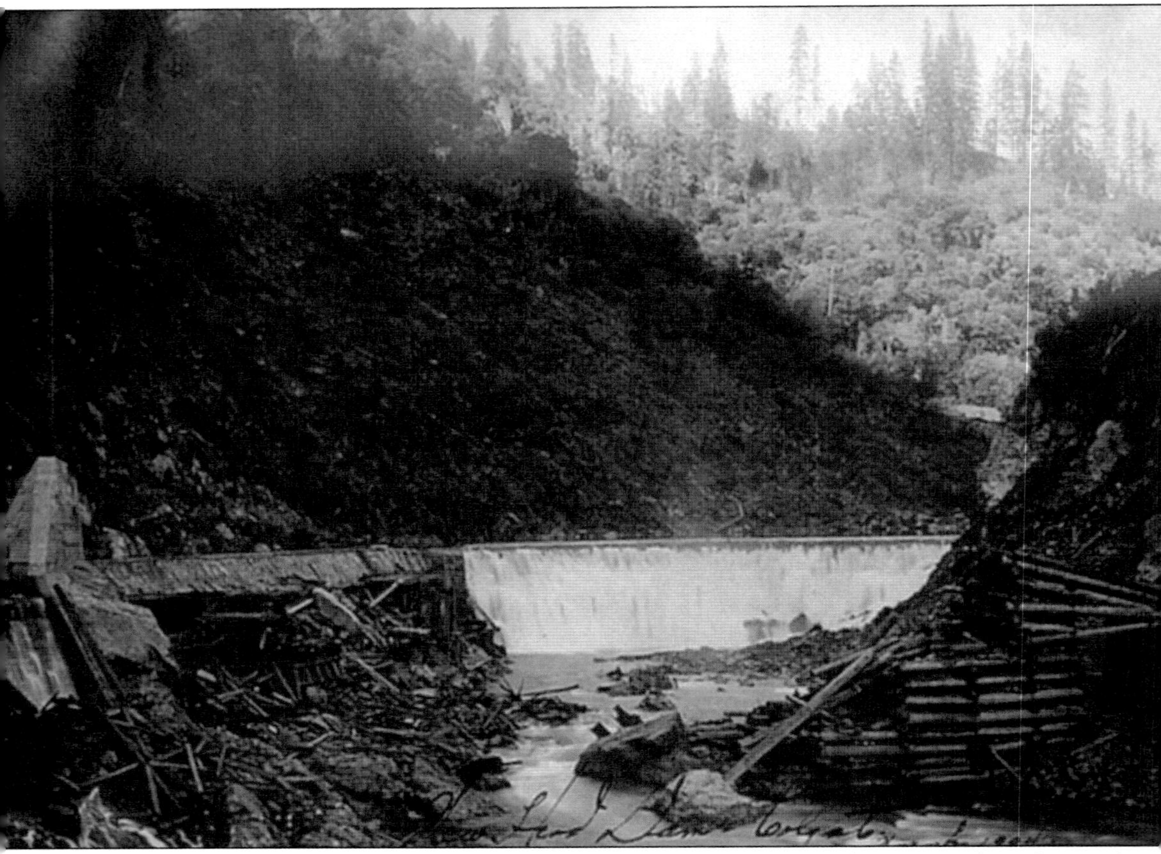

Several dams have served the Brown's Valley Irrigation District over the years. When the need for more water arose, an agreement was struck with the Bay Counties Power Company to build the new Colgate Dam. It began operation in 1904, and could generate electricity as well as supply the quantity of water necessary for the Brown's Valley Irrigation District's purposes. The original crib dam can be seen here, broken in the foreground, with the new rock and concrete dam behind it. (Courtesy of Dale Johnson.)

Six

Dredging
The New Big Business

When orchardist Wendell P. Hammon decided to dig a pit in some tailings near the Feather River to test them, he found that great quantities of gold remained within. People had begun dredging for gold in New Zealand in 1865, and it was thought that the process would work well in the streams of California. It did for a while, but the dredgers were not able to work the gravel thoroughly enough to make it pay. If a new method to process the gravel faster and more efficiently could be developed, it would yield a fortune. An article in *Scientific American* gave Hammon the idea to improve the dredging process with a continuous bucket-line design.

In 1898, Hammon went into the dredging business with investor Thomas Couch, and their Yuba Manufacturing Company built the first of many successful California dredgers. By 1909, dredging had become the new big business and Hammon was the head of three major dredging operations in California. One of the three, Yuba Consolidated Goldfields, controlled most of the acreage in Yuba County. Innovations made on the goldfields allowed dredge designs to continually evolve and improve, and by 1912, dredgers could dig to a depth of 80 feet and had buckets with a capacity of 15 cubic feet. The dredges were becoming so efficient that even areas with lower gold concentrations could be worked at a profit. Just as the Yuba gravel lands appeared to have been worked out, the price of gold rose and further improvements were made to the equipment, allowing the dredge to dig to a depth of 110 feet—all the way to bedrock. Yuba Consolidated Goldfields became the industry leader, and Yuba County was once again leading the world in gold production. Dredging provided jobs for many men who had worked the earlier types of mining—some of whom were the third-generation descendants of the original forty-niners—and brought new prosperity to the mining towns of Yuba. Gold was harvested this way in Yuba for at least another 50 years.

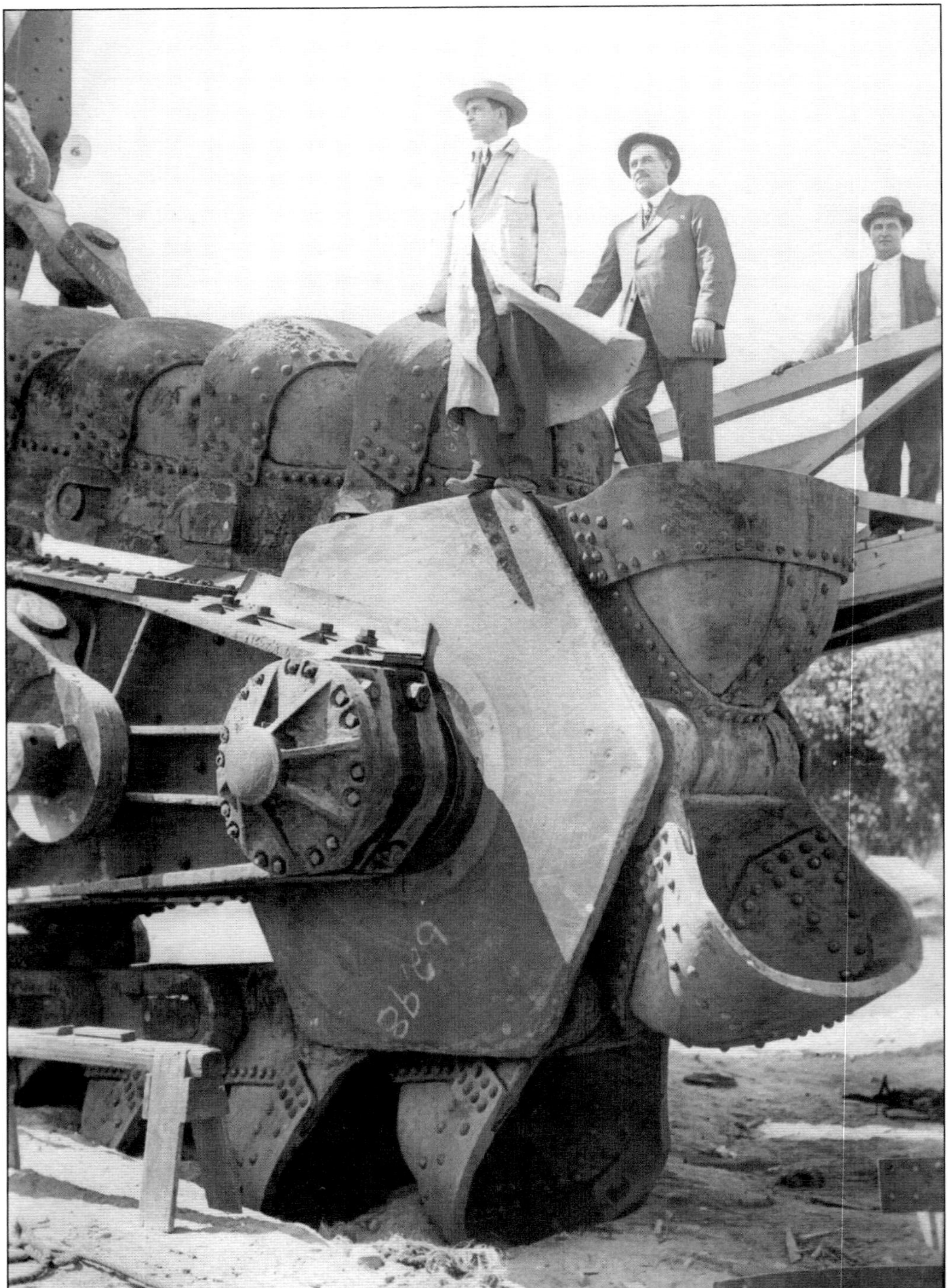

The fortuitous reading of an article in *Scientific American* gave W.P. Hammon the idea of using a continuous bucket line dredge. The original machines were small and could only work 1,000 yards a day, but over time, they evolved into the giant machine pictured here—which could work over 1,000 times more material. Newton Cleveland is at left, with Hammon standing at his right. The man on the walkway at far right is unidentified. (Courtesy of Yuba County Library.)

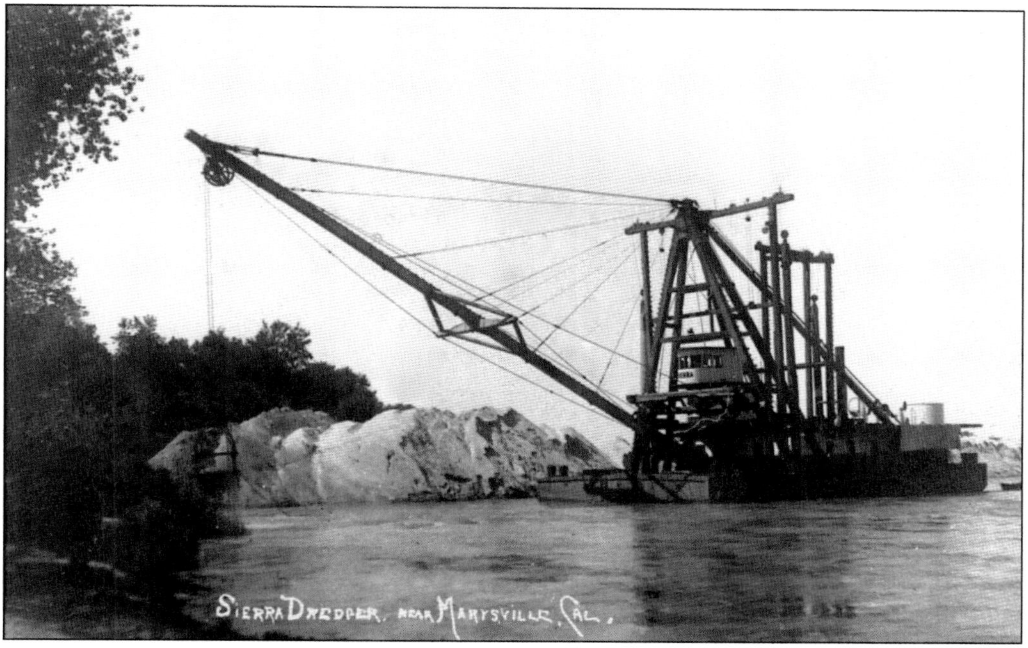

Gold dredging in California began with a small riverboat that was fitted out as a dredge similar to a steam shovel, and dredging was attempted on the Yuba River. This was a good idea, as the dredge could have helped clear the river, but it did not have the capability to wash the gravel on board. Single bucket dredges like this did not produce enough value to sustain their operations. (Courtesy of Yuba Feather Museum.)

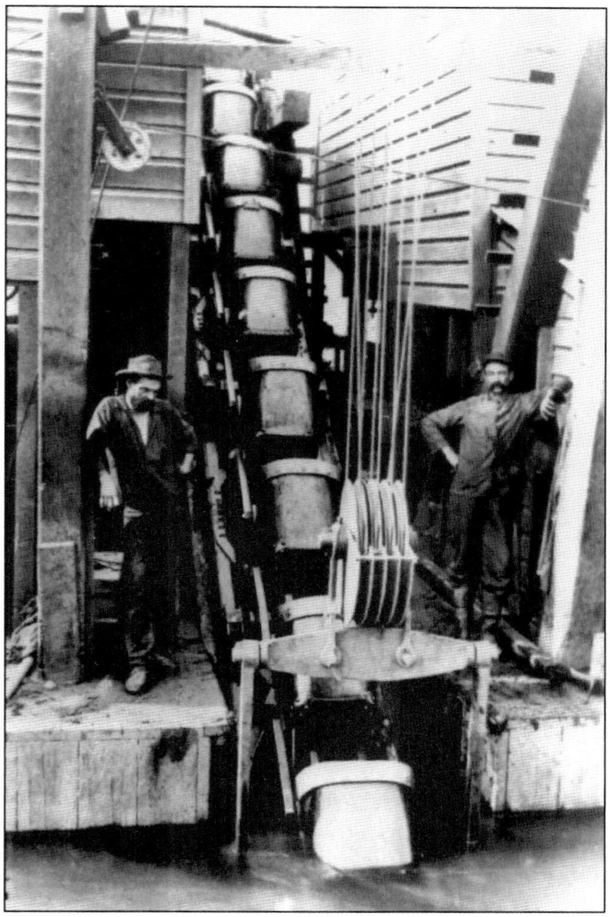

Pictured here is a continuous bucket line. It is bucket after bucket in a continuous line, which allowed a greater quantity of gravel to be brought up and processed on board the dredger. (Courtesy Ruth and Doug Criddle.)

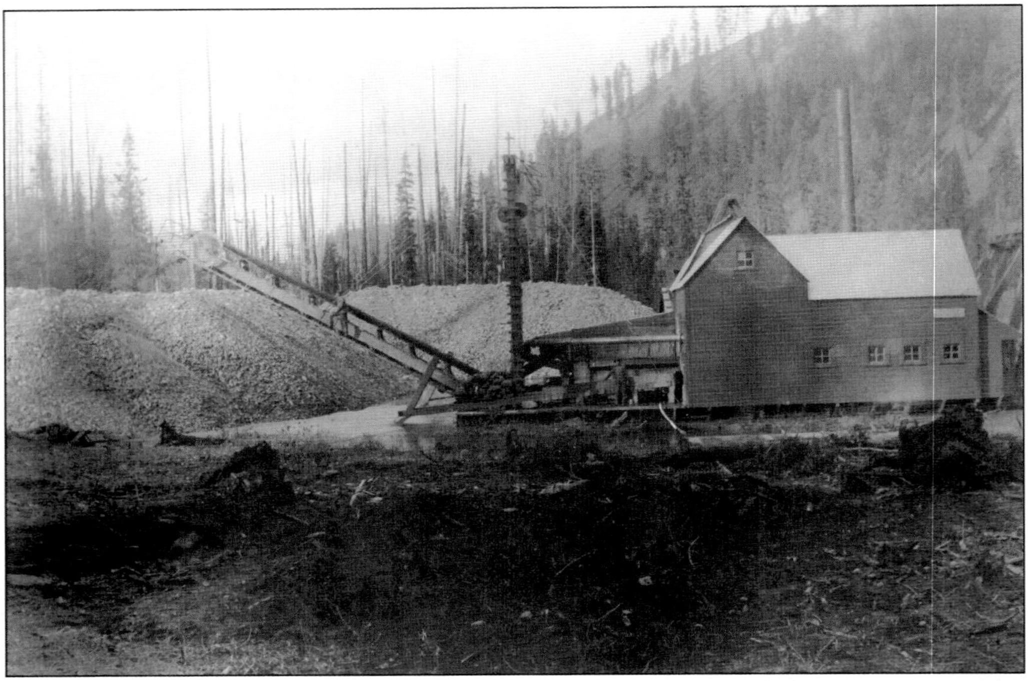

The first successful California gold dredge was built in 1898 by Wendell P. Hammon and his partner, Thomas Couch, a Montana mining businessman. Hammon founded the Yuba Construction Company in 1909 to build dredgers. The components were built at a shop in Marysville. The first dredger in Hammonton began operation on August 1, 1904. (Above, courtesy of the Marple/Hapgood family; below, courtesy of Yuba County Library.)

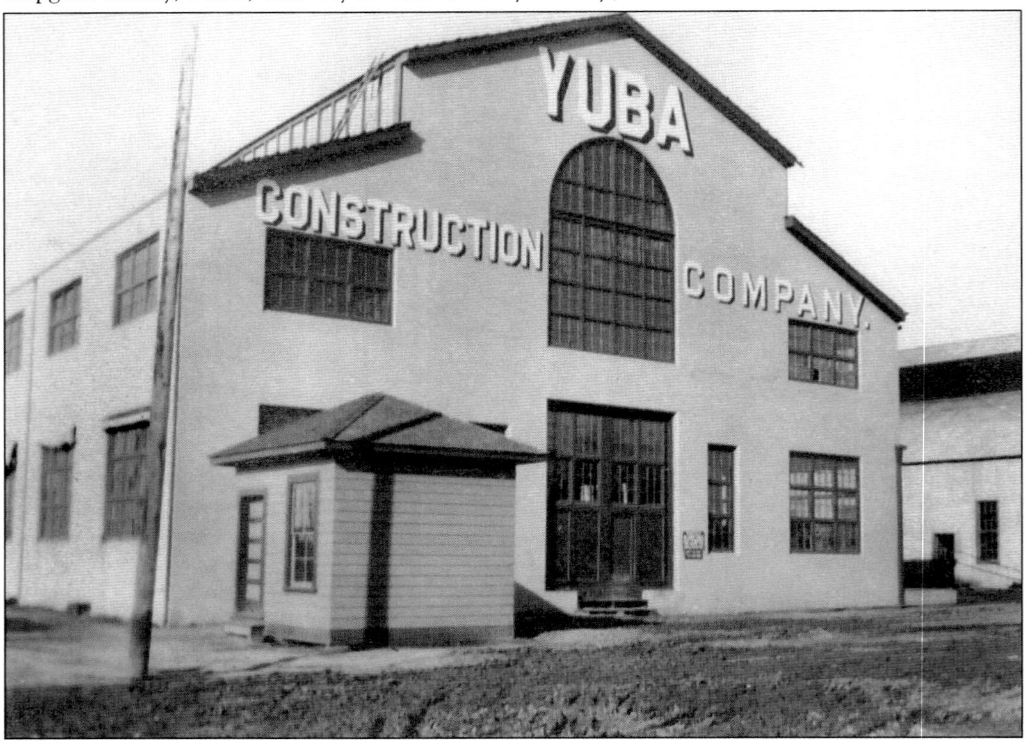

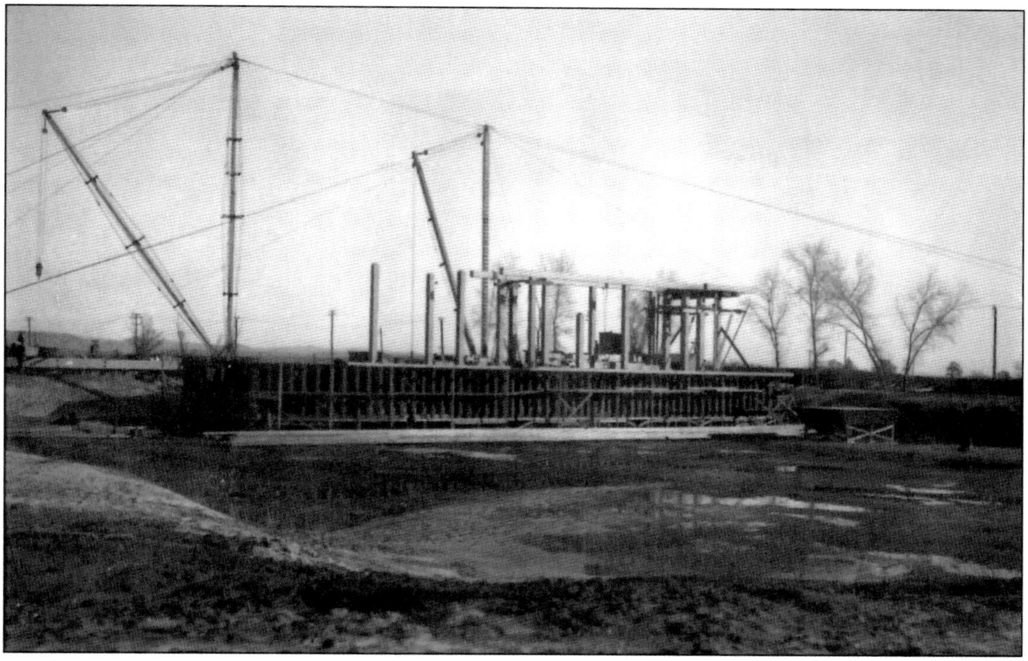

Dredgers were assembled on the spot in Hammonton. The large components were hauled there using the Yuba Tractor, which was also manufactured by the Yuba Construction Company. The first dredge was powered by burning 15 cords of wood a day. John Martin and Eugene deSable were just beginning to supply electricity to Marysville and convinced Hammon to try electricity on the dredges, at their expense. Their gamble paid off, and all of the following dredgers were powered by electricity. (Both, courtesy of Ruth and Doug Criddle.)

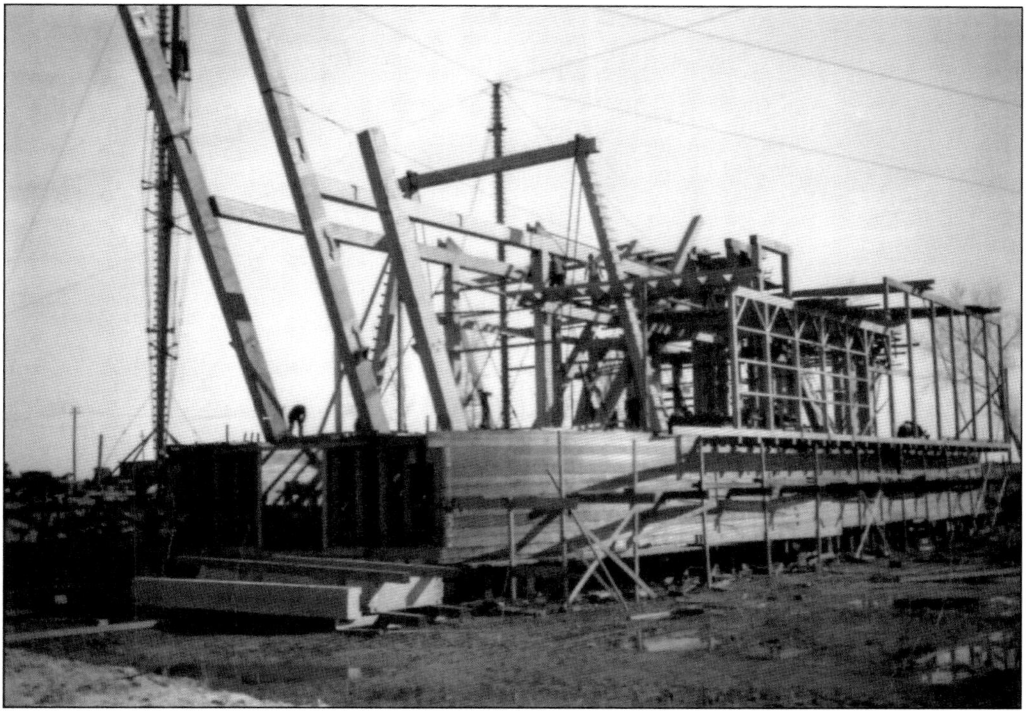

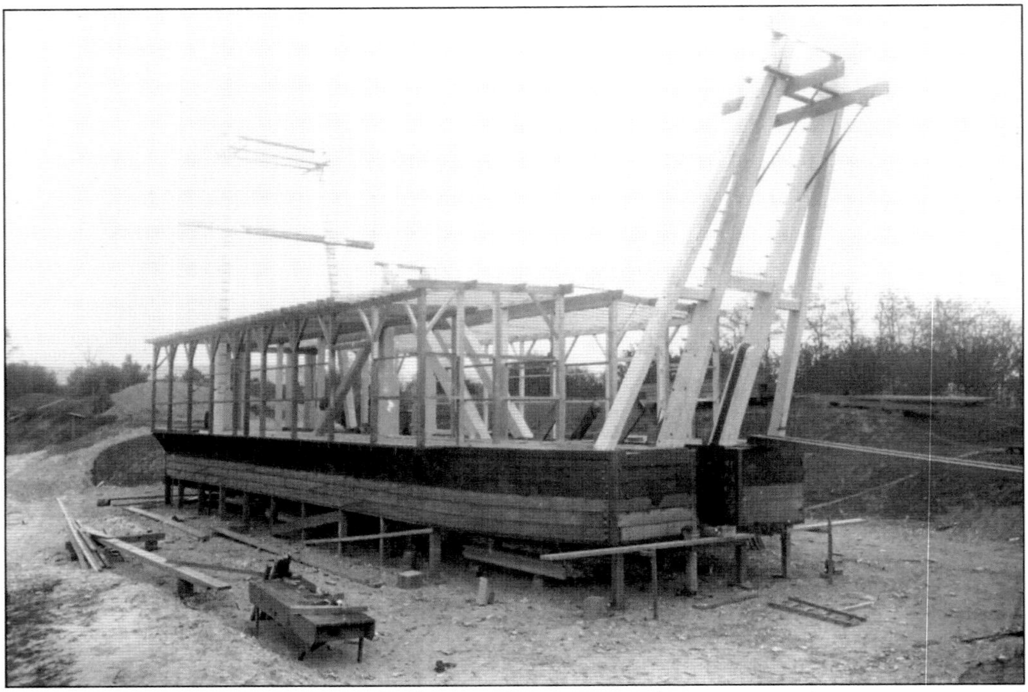

Dredgers float in water, but these fields were dry. Pits were dug on site and water was pumped to allow the dredger to float while it worked the gravel. This opened up vast areas to dredging work. As the buckets dug into the bank, the dredger was propelled forward. (Courtesy of Ruth and Doug Criddle.)

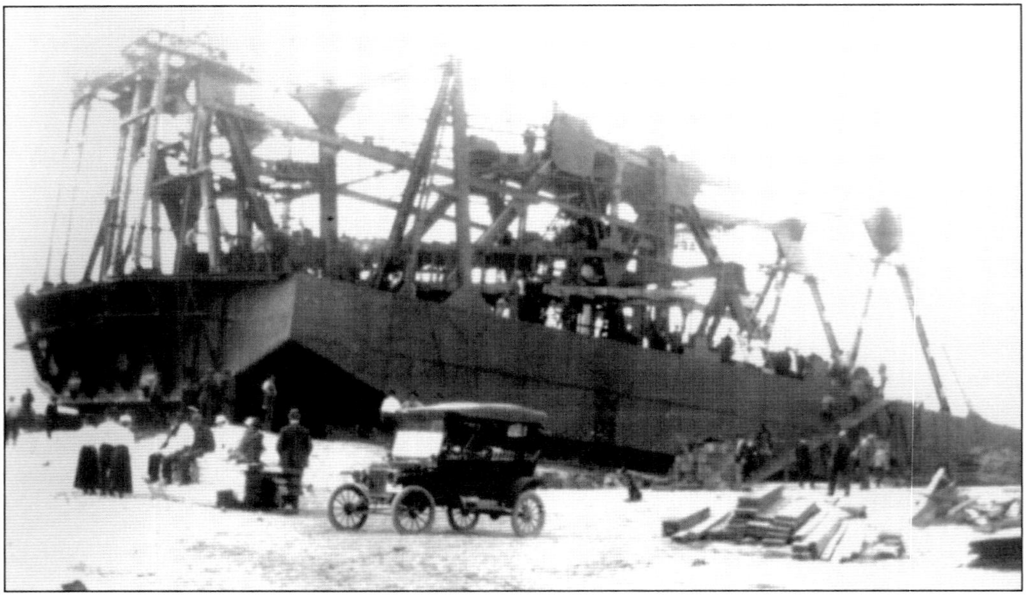

In 1905, the state mineralogist listed 10 dredgers—all supplied by Yuba River water—that worked an average of 3,000 cubic yards per day. They were powered by electricity generated onsite, and eight belonged to the Yuba Consolidated Goldfields Dredging Company. The other two belonged to the Marysville Gold Dredging Company, which was also called Marigold. The owner and general manager of all 10 dredgers was listed as W.P. Hammon. (Courtesy of Ruth and Doug Criddle.)

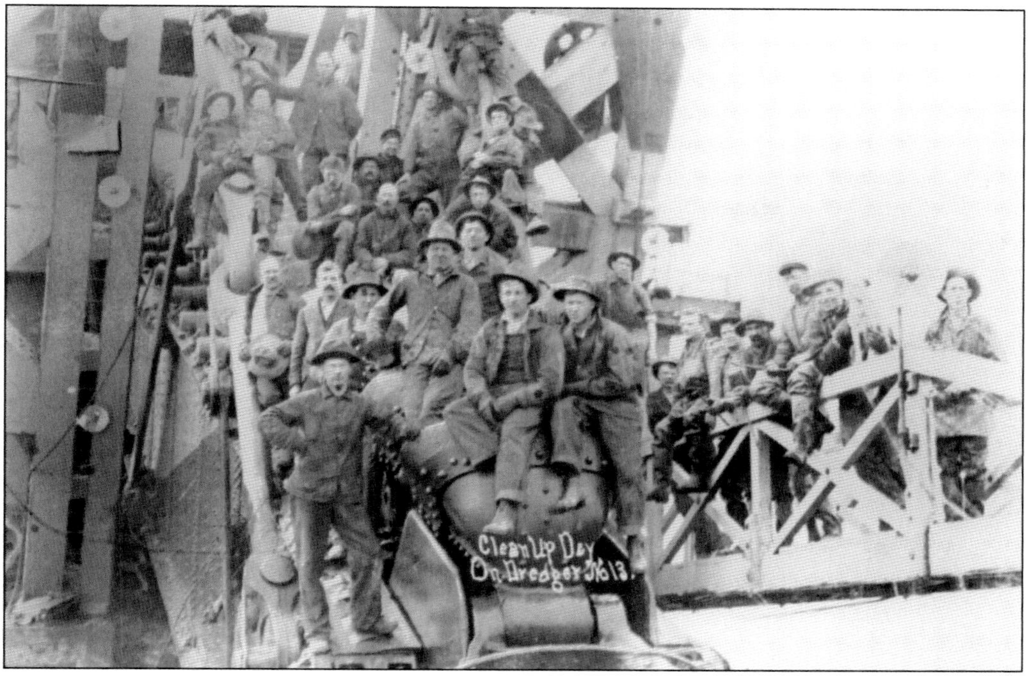

Dredger No. 13 was one of the last wood dredgers and began operation in 1911. After 10 years in operation, No. 13 had exhausted all dredge-able ground in its vicinity and produced receipts for $1,417,847.15 worth of gold. (Courtesy of Ruth and Doug Criddle.)

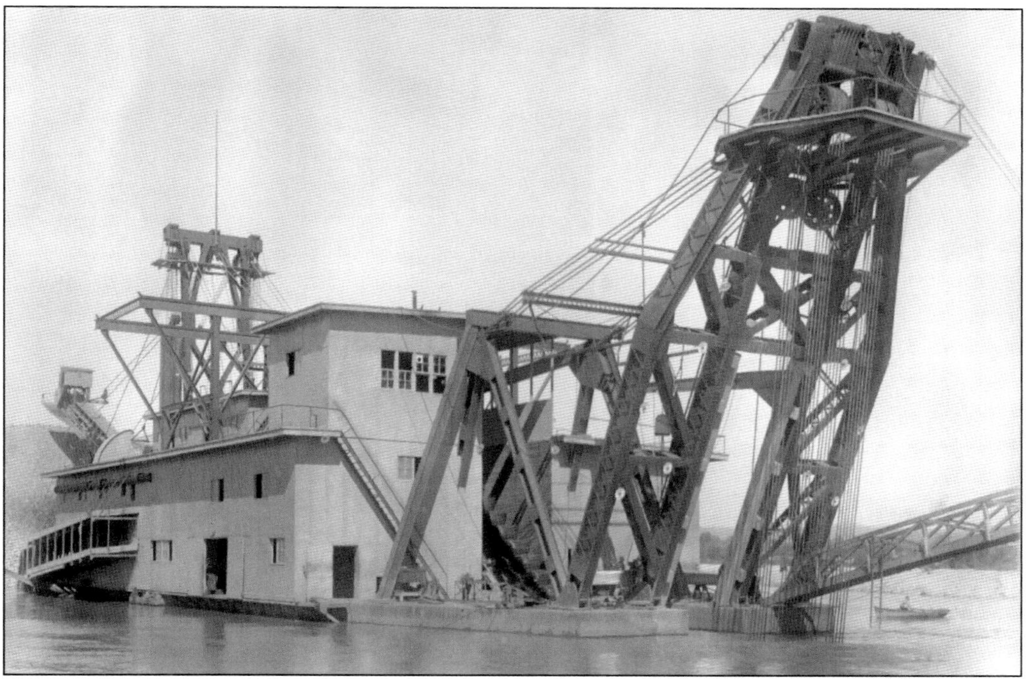

Steel came to replace the earlier wood construction of the dredgers. Some of the components of a dredger were a floating hull, a digging ladder, an endless chain of buckets, a screening apparatus, gold-saving devices, pumps, and a stacker. (Courtesy of the Yuba County Library.)

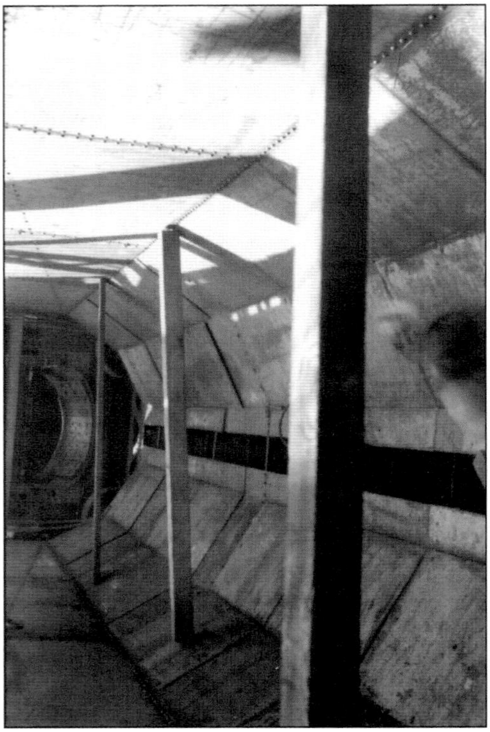

A dredger separates the gold in a process similar to older methods of washing gravel. The dredger digs into the gravel and sand, and when the buckets come up with the ore, it is separated by running it through the trammel—a cylindrical device that revolves to separate and discard rocks and larger gravel chunks. Further processing of the dirt and sand uses a system that is also similar to older methods. (Both, courtesy of Ruth and Doug Criddle.)

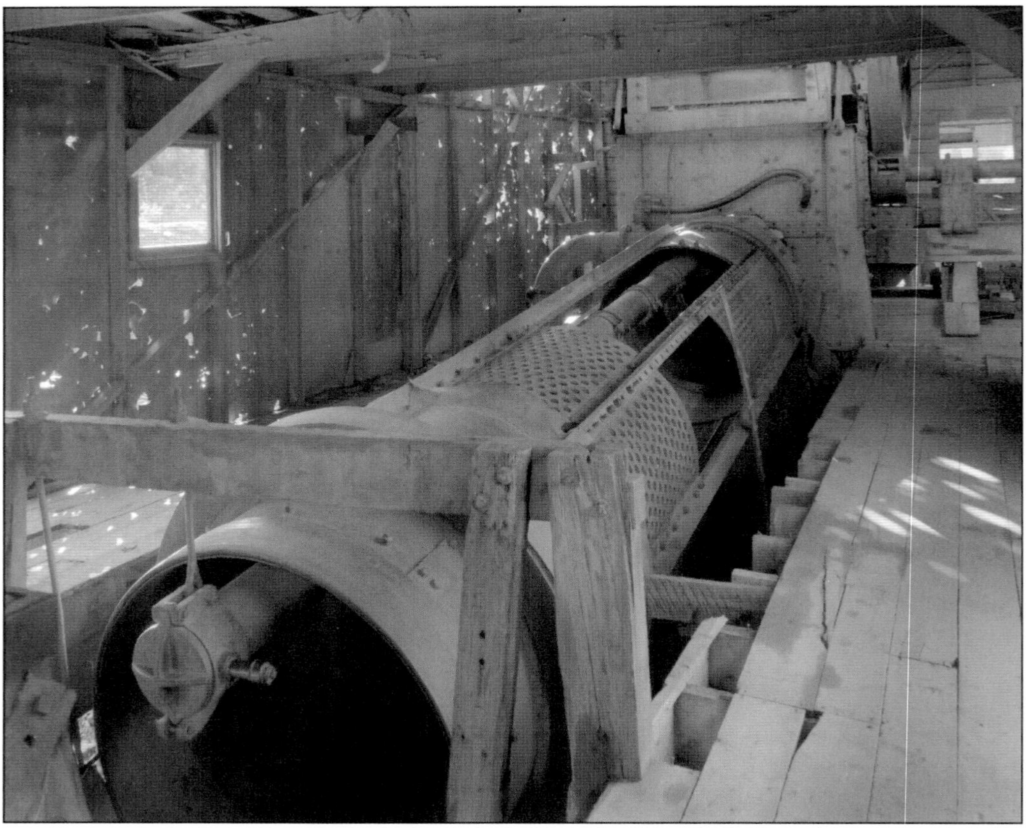

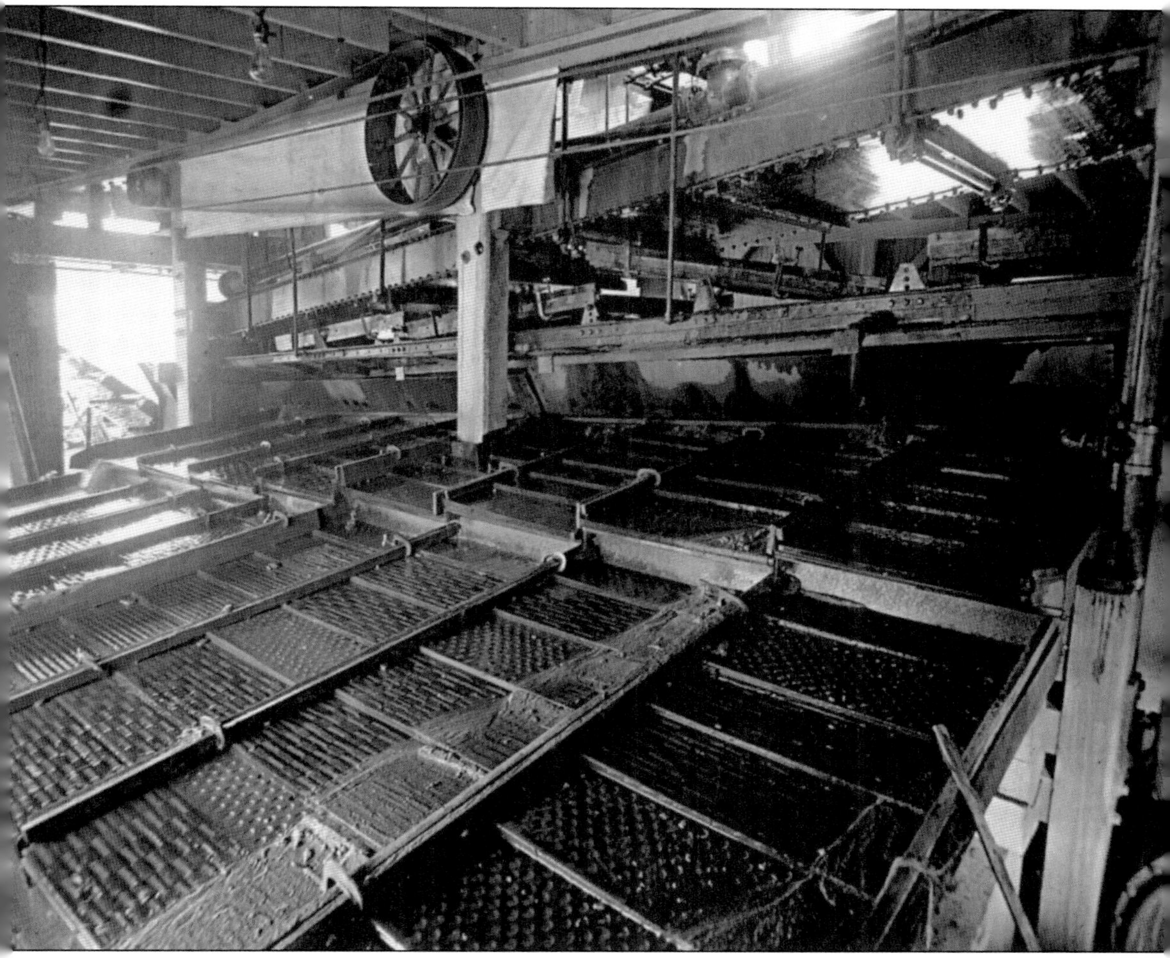

These are the raffles that process the sand and dirt. Ore is washed through them, allowing the heavier gold to fall down and the lighter sand and dirt to wash out. These more modern raffles were made of a rubber-like substance. Rather than running the tailings over a long stretch, they were kept in a smaller, confined space and agitated—similar to an early rocker. (Courtesy of Yuba County Library.)

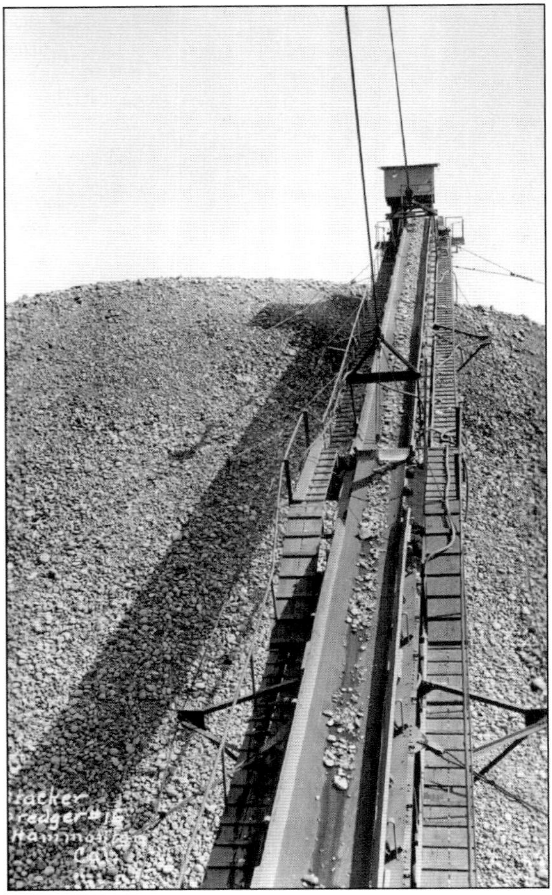

This is a detailed image of a stacker. After processing, the discarded dredging debris is transported by the stacker and piled behind the dredger, forming unique patterns. (Courtesy of Ruth and Doug Criddle.)

These unusual mounds are left behind by the stackers as the dredger moves through its work area. Their patterns vary based on the dredger that is leaving them. Some have only one stacker, while others have several. However the pattern varies, it is a clear sign the area has been worked. (Courtesy of Yuba County Library.)

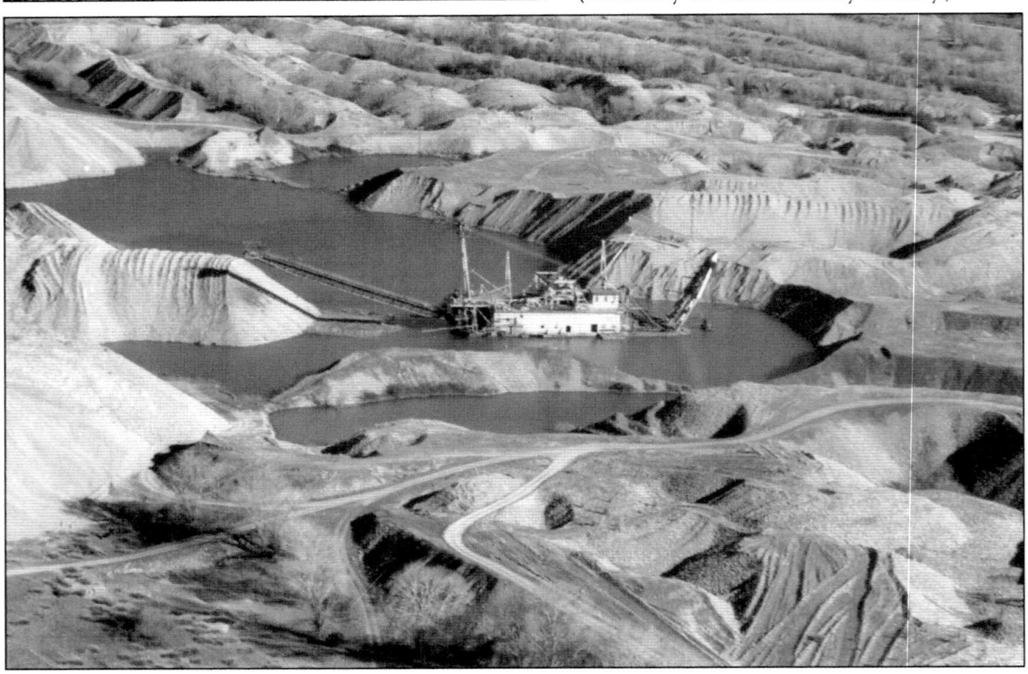

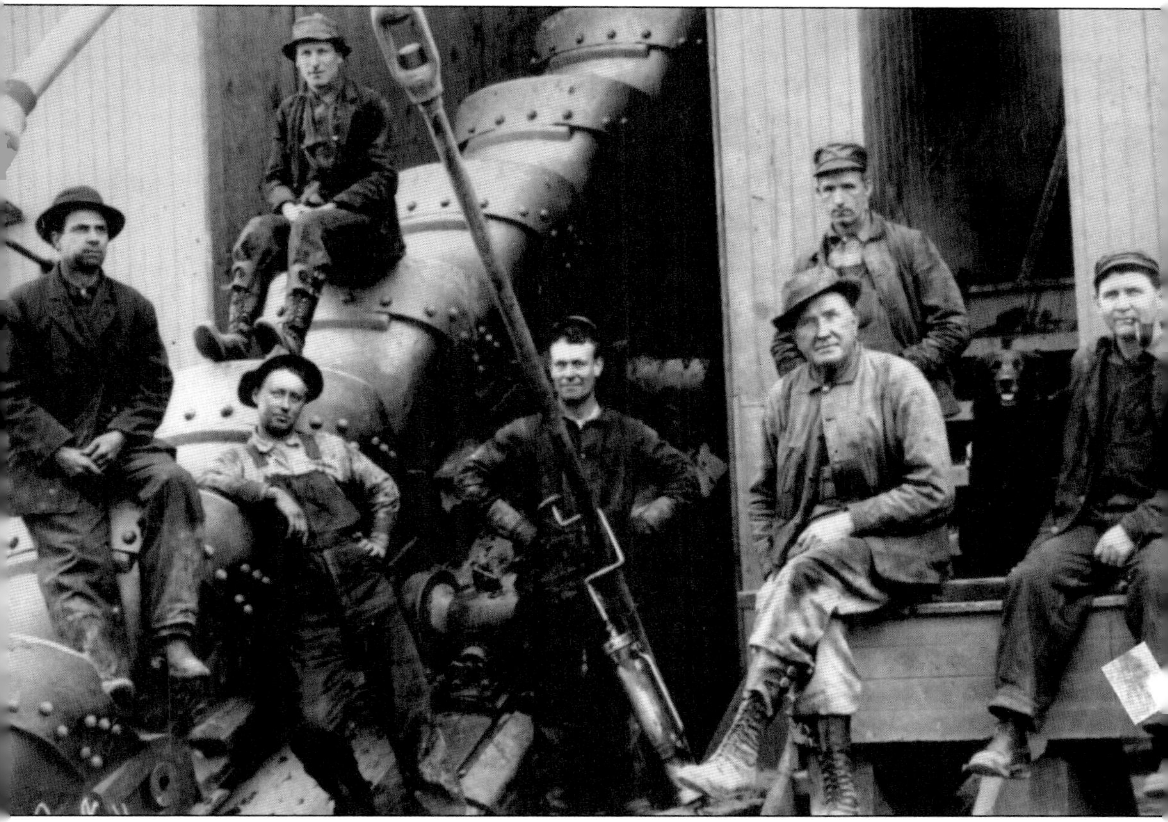

Dredgers had a crew of between 8 and 15 people per shift, and they worked in shifts around the clock. This photograph shows a portion of the crew on Dredger No. 11. The man sitting at the top of the bucket line is Emmit Schofield, and the man directly below him is Charlie Arnold. The man with his legs crossed at right is identified only as Arnold. The others are all unidentified. (Courtesy of Ruth and Doug Criddle.)

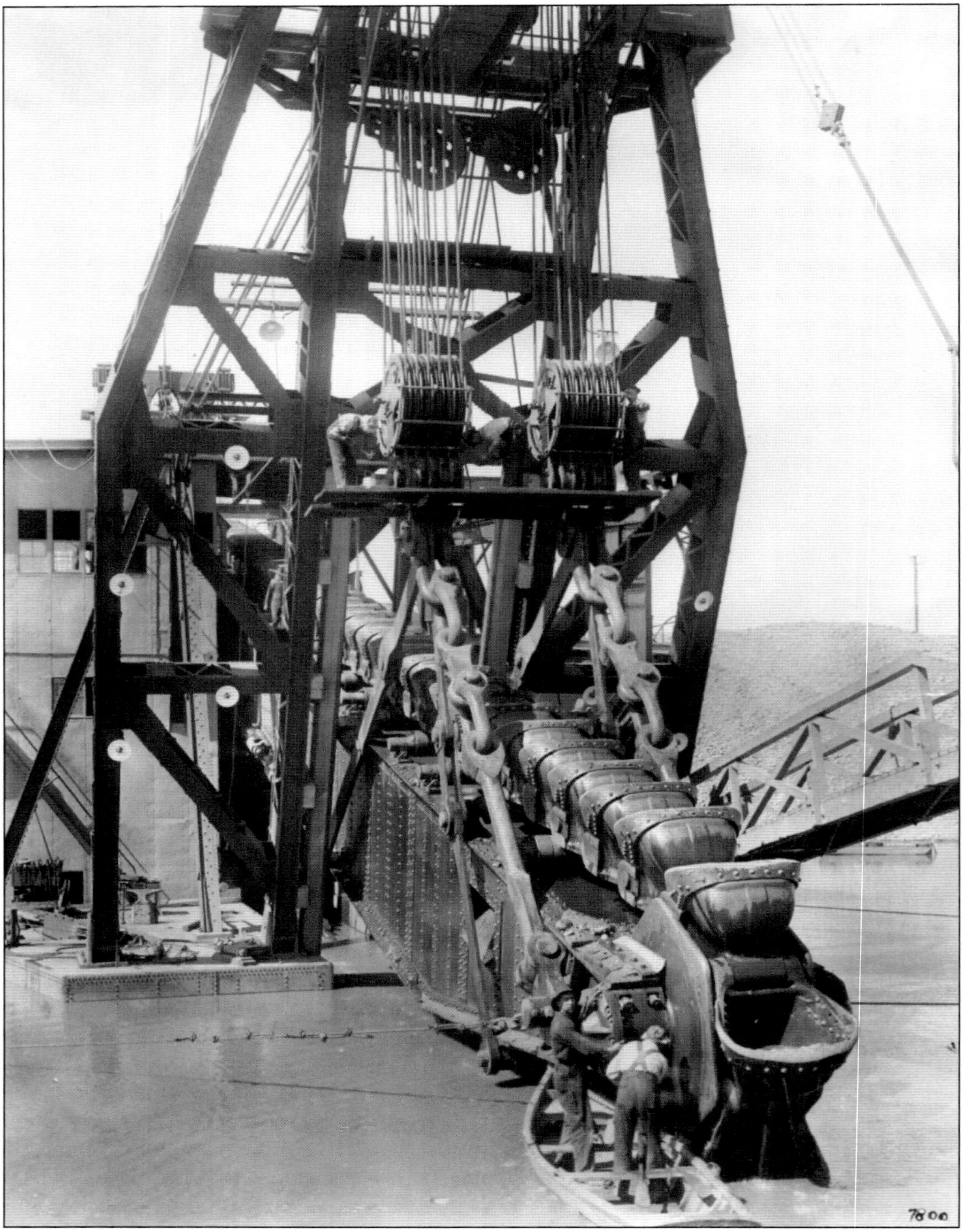
The dredging operations were a large employer in Yuba County. They hired not only the crews that ran the dredgers, but also the many who worked behind the scenes in the shops rebuilding, maintaining, and repairing the equipment. Maintenance was a large part of the job. Here, some men in a rowboat appear to be repairing the bucket line. (Courtesy of Yuba County Library.)

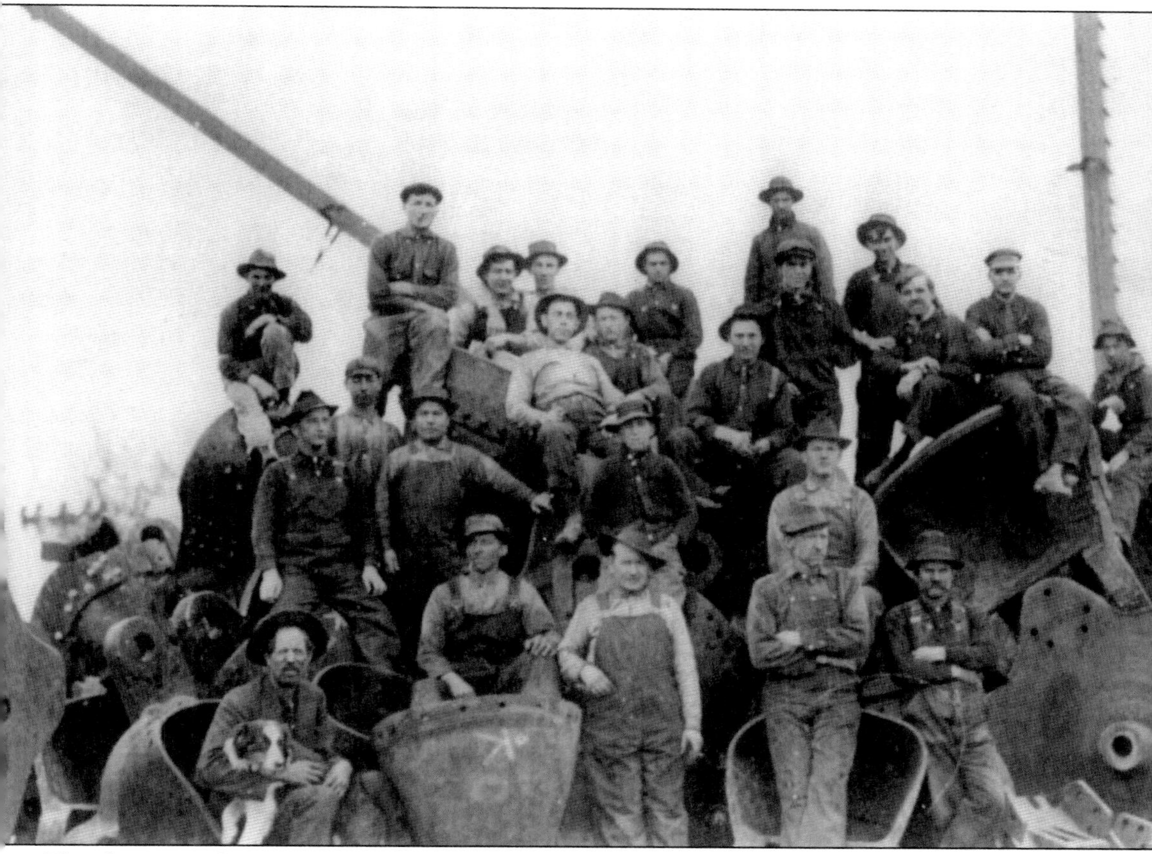

The shop crew from Hammonton posed here with various dredger pieces and parts. Included in the photograph are Ed Carlin, John Copeland, Doc Kendall, L.E. Tipton, Sandy Vand, Perry Frederick, Gordon Rhyne, Joe George Walk, Jim Ludd, Lloyd Readdick, and Bill Readdick. (Courtesy of Ruth and Doug Criddle.)

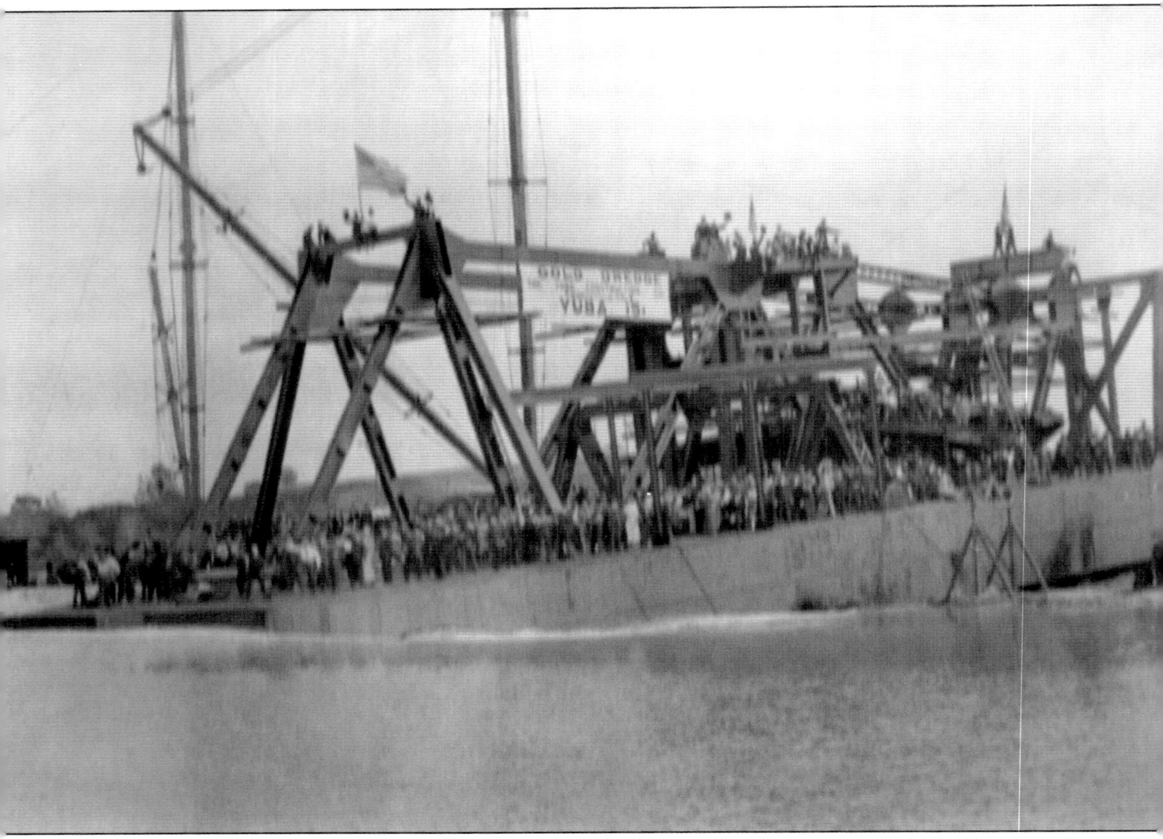

A huge celebration was held for the launch of Dredger No. 15 on July 11, 1816. This must have been an exciting event, as nearly every resident of the town of Hammonton was on board. Dredger No. 15, along with all mining activity in the area, was shut down during World War II. In 1942, the War Production Board Limitation Order declared that gold, which had been a determining factor in the outcome of the Civil War, was not necessary to the war effort and that the miners should be working in industries that dedicated all available resources to the war effort. Local mining never really recovered, but this dredger resumed operation in 1945 and functioned for another 10 years. Dredging operations fared better than mining as a whole. (Courtesy of Ruth and Doug Criddle.)

Seven

There Is Still Gold in Them There Hills

Will it Stay There?

There is still gold concealed in these hills, but there are only a few working mines left in Yuba County. The gold will probably remain there unless some future technology makes it ecologically safe and economical to get to it. While there are many weekend prospectors and recreational gold hunters out there, regulations are rigorous and legal extraction methods must be employed.

Today, the private ditches that were built to supply water to mines are public irrigation districts and are used for domestic water and agricultural irrigation. The dams that originally restrained tailings have been augmented for hydroelectric power generation and water storage. Much of the land that was mined stands largely unused and unchanged from 100 years ago. Some has been used for cattle grazing, some is used for other agricultural purposes, some stands as a wildlife refuge, and even more is used by the University of California for research purposes. A large portion has also been preserved for recreation and educational purposes.

Environmental groups have been successful in mitigating some of the damage to the Yuba River by restoring some salmon spawning grounds, but there is still much to be done. Mercury is hidden deep in the tailings, and researchers are working to determine how much there is, where it is, and how to clean it up without causing more damage. Some groups advocate taking down the dams so that fish can return to their ancestral spawning grounds, while others want to build more dams for water storage and energy production. A by-product of mining, the sand and gravel business, thrives by tearing down hills for the materials used in the constant construction of roads and highways. Only time will tell what lies in store for the mining lands in Yuba.

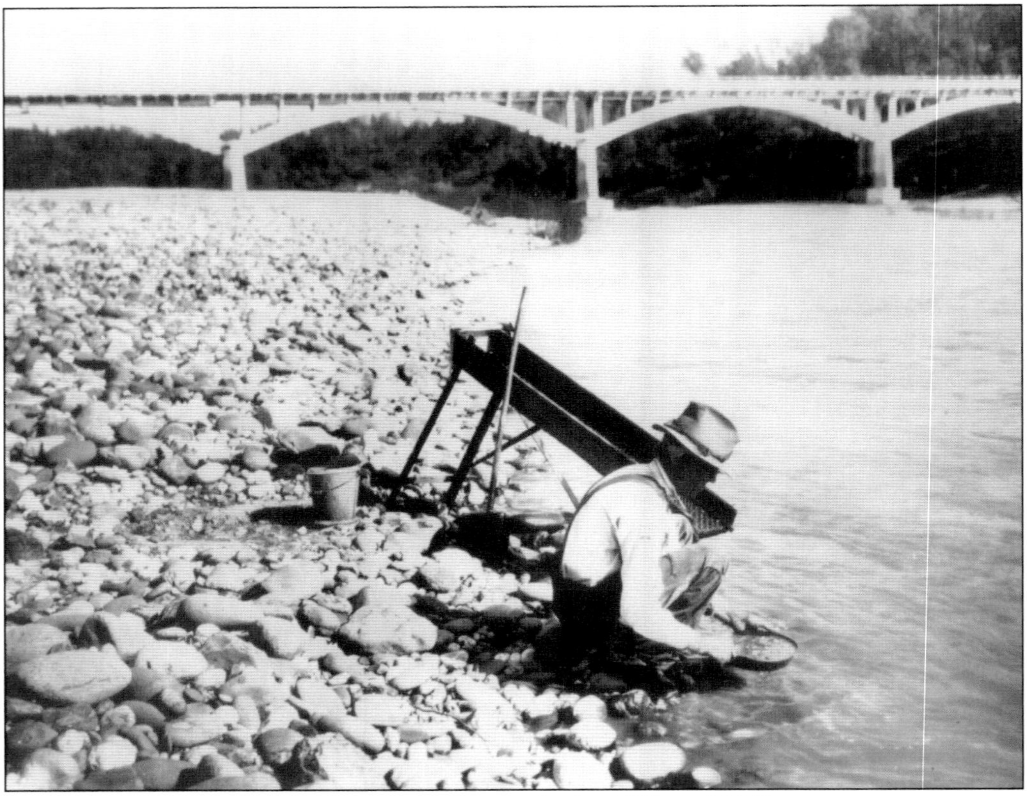

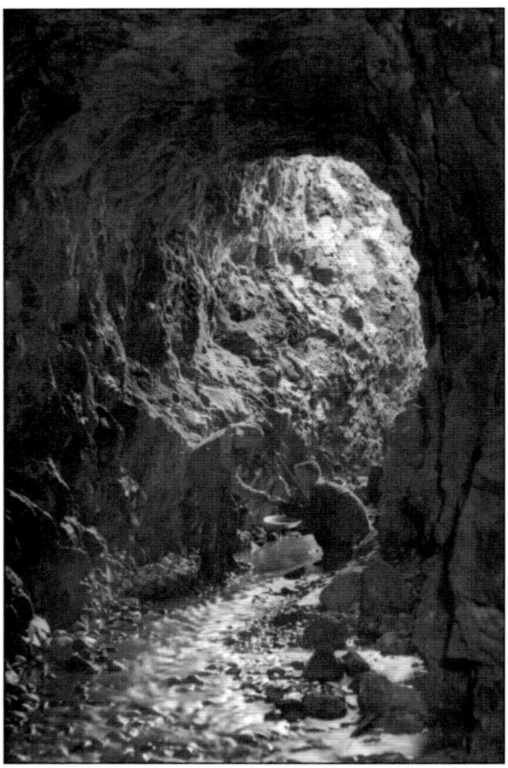

In the 1930s and 1940s, many came to the banks of the Yuba River—refugees of the Great Depression and the Dust Bowl. Much like the prospectors of 1849, these refugees hoped to make enough from a little river claim to keep their families fed and together. Although there were some WPA projects in the area employing men to build the Upper Narrows Dam, there were still many who prospected on the river for some financial assistance. Many long time residents of the area said that they only made it through the Great Depression by doing some panning or some crevice mining in the old tunnels. If they were lucky, they could get enough to buy some necessities. Pictured at left, at the end of a tunnel, Cecil French (left) passes on the skill and story of panning to his son, Bob. (Above, author's collection; left, courtesy of Charles French.)

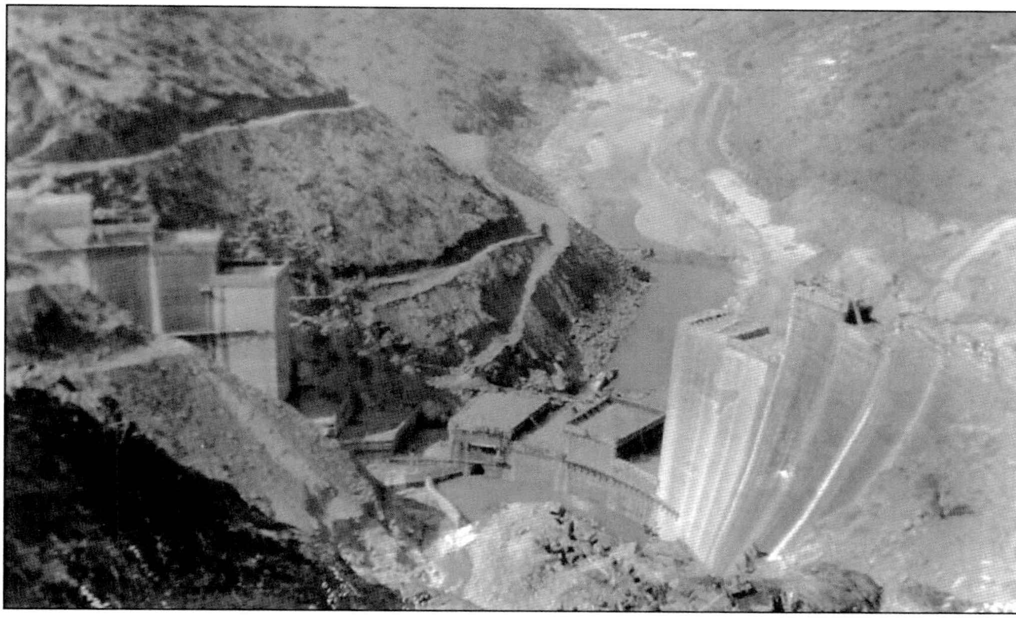

During the Great Depression, the price of gold increased to $35 per ounce. Following the stock market crash of 1929, people began to look at hydraulic mining as a way to revive the economy. Congress allocated money to build dams, including the Upper Narrows Dam, to control debris and restore the mining industry. The completed reservoir and dam were then named for Harry Lane Englebright, the congressman from Nevada County who championed the cause. Construction begun in 1938 and was completed in 1941 at a cost of $4 million. Originally built to catch hydraulic mining debris, the dam was later retrofitted for hydroelectric generation. The dam is 260 feet high and 1,142 feet long, with a reservoir that is 227 feet deep at the dam, covers 815 acres, and is 9 miles long with 24 miles of shoreline. (Above, courtesy of Rosemary Freeland.; below, courtesy of Leanna Poe Beam.)

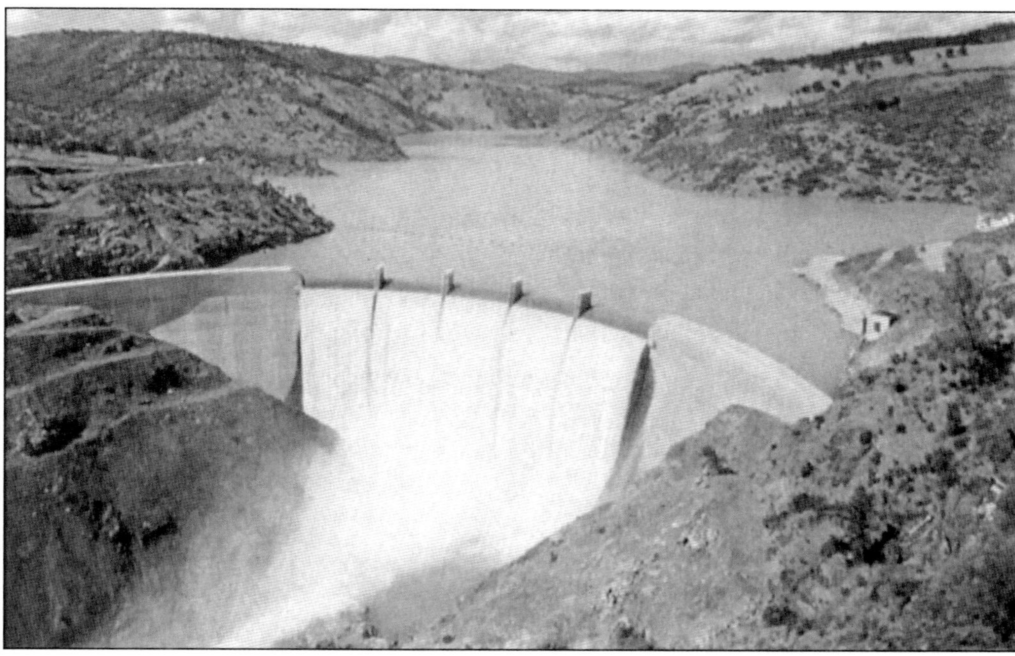

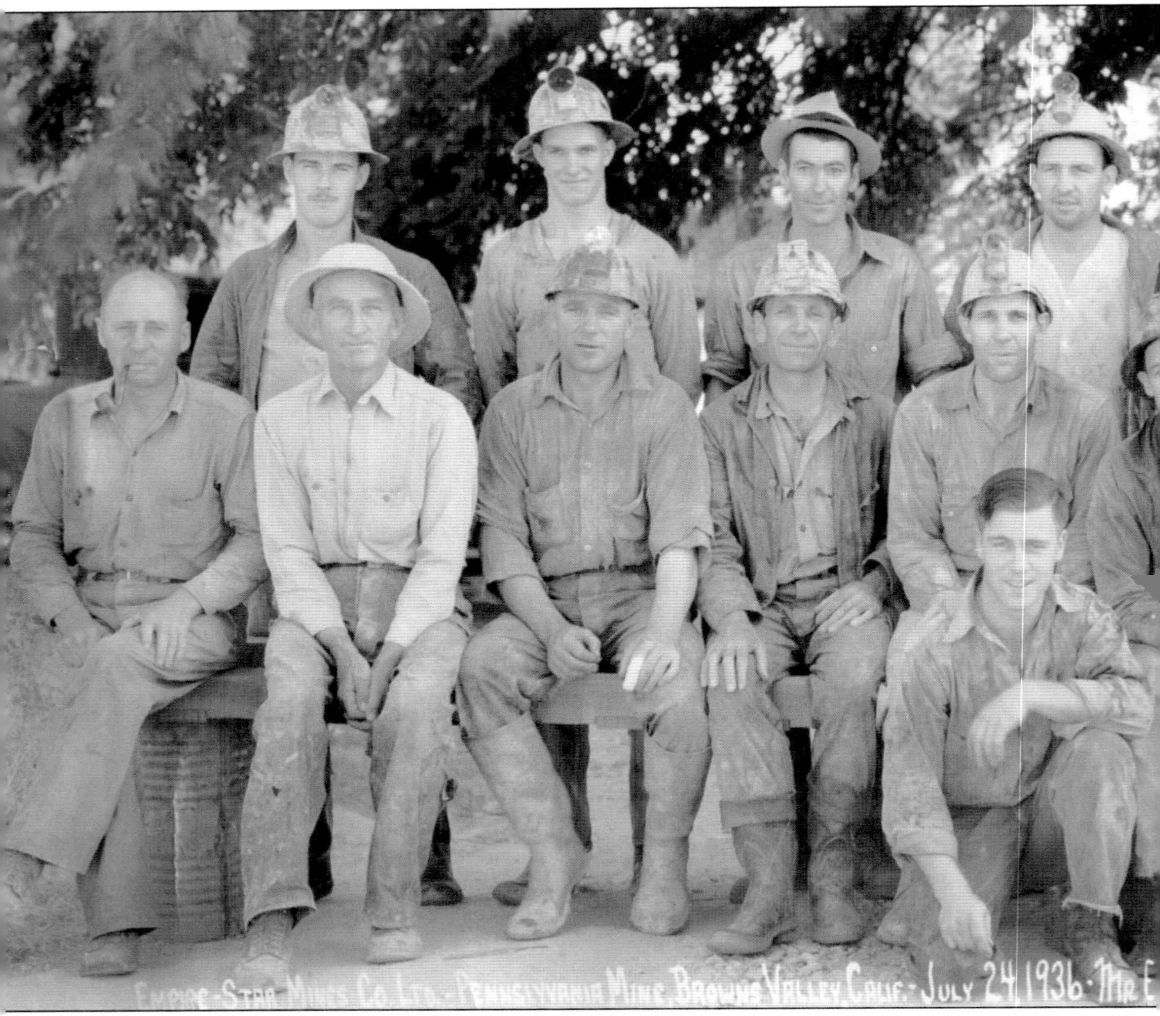

The Pennsylvania Mine operated for a long time. It was worked by several different companies and, in 1936, was part of Nevada County's Empire Star Mines. This group portrait provides an interesting comparison to the one on page 36 showing the men who worked the same mine in

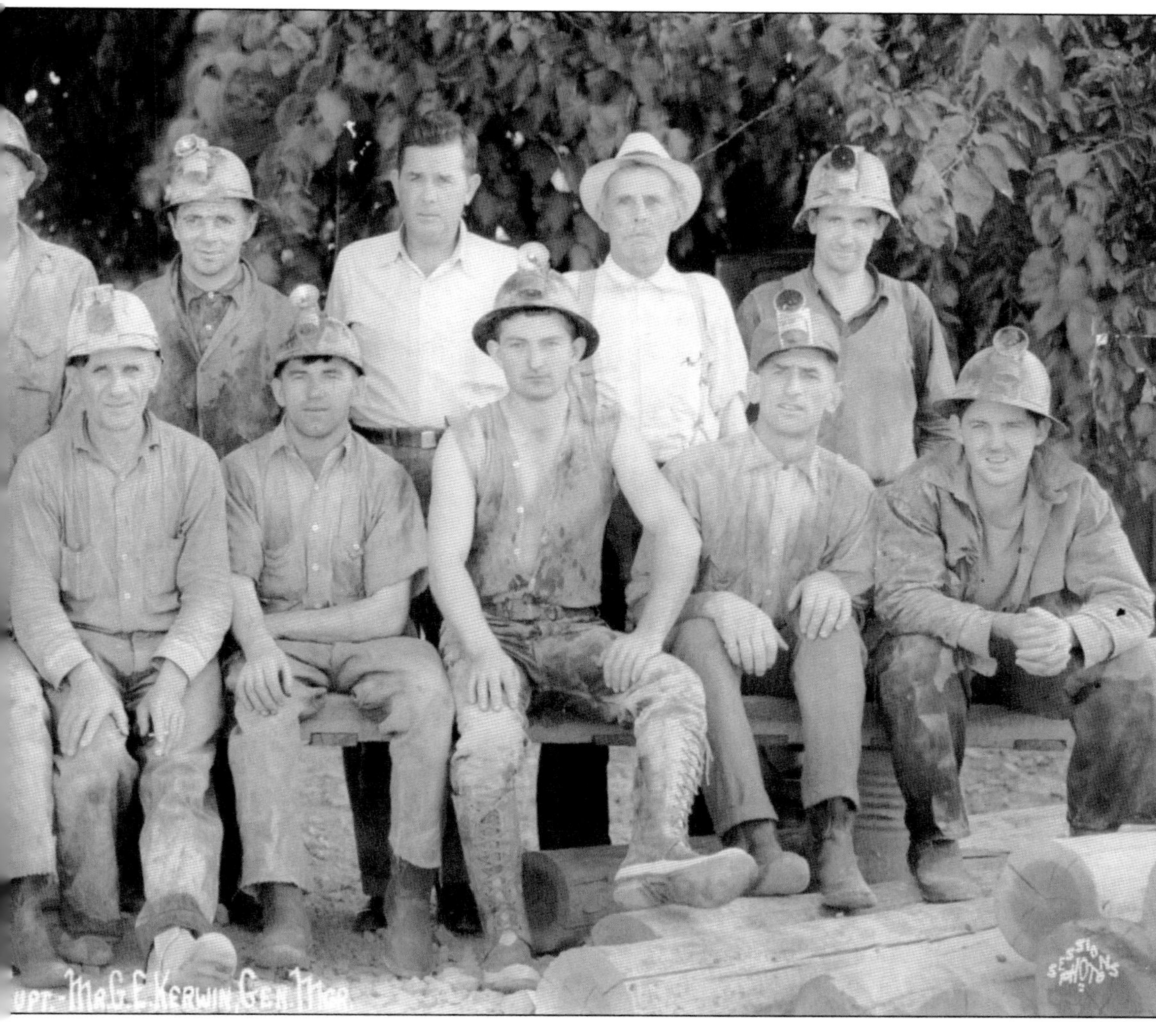

1902. This image is certainly clearer, but the look of the miners is quite similar. (Courtesy of Roberta D'Arcy.)

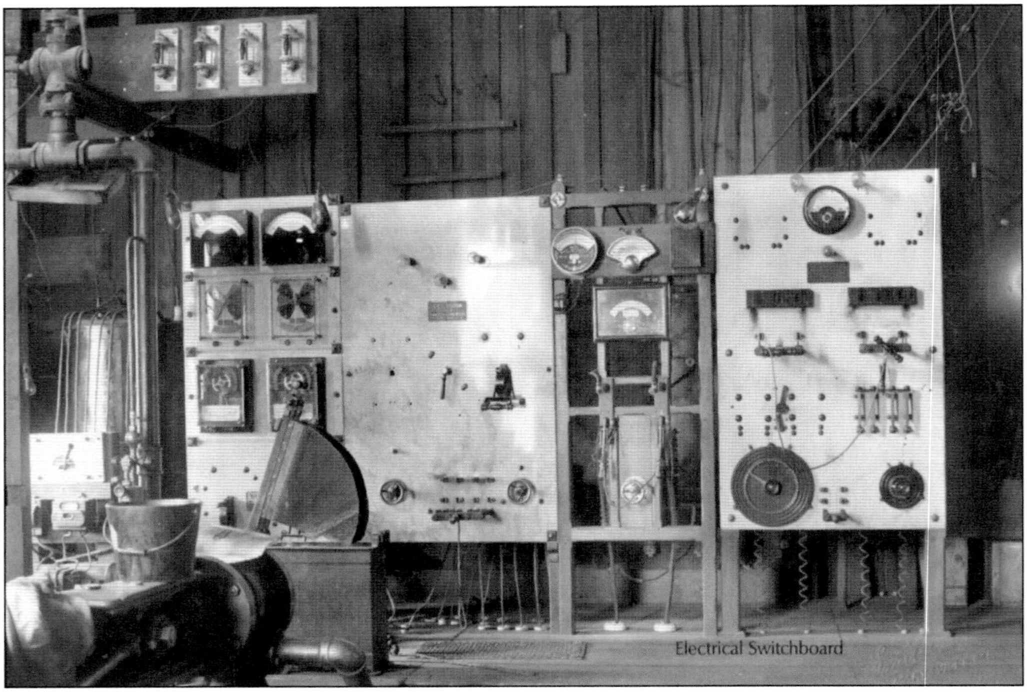

The Brown's Valley mines were worked longer than many others. Over time, different companies operated them. With each change in management, new ideas and fresh capital brought improvements. As new technologies became available, the mines used them to their advantage. Pictured here are the electrical panel (above) and engine room (below) that generated the power to operate this new equipment, the hoist, and the ore cars. (Both, courtesy of Ivadene Leech.)

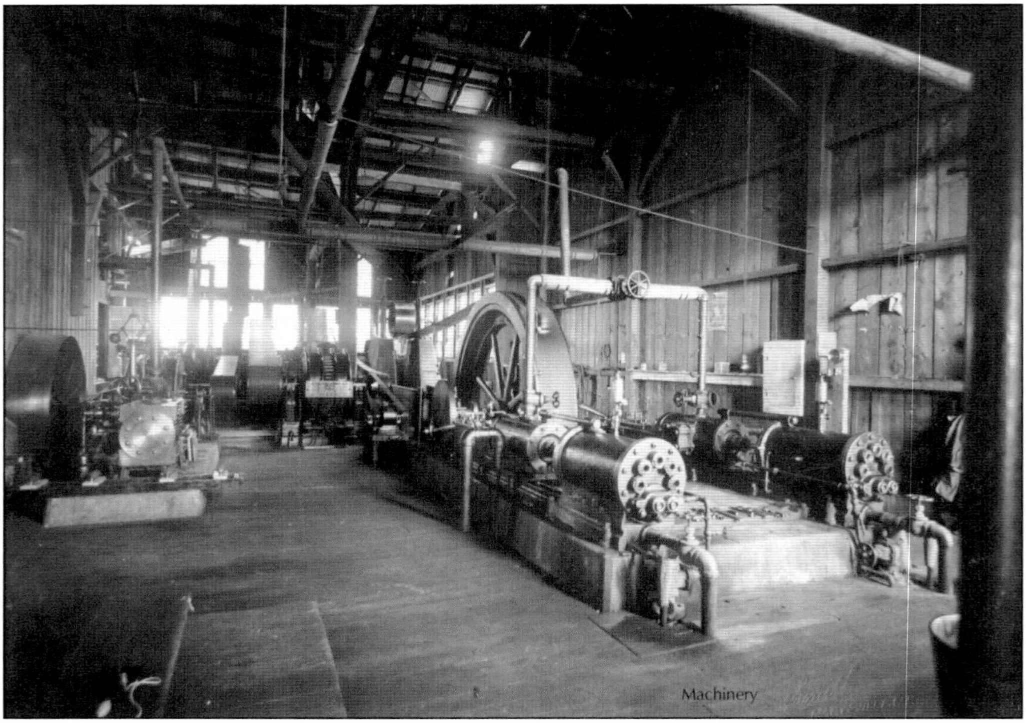

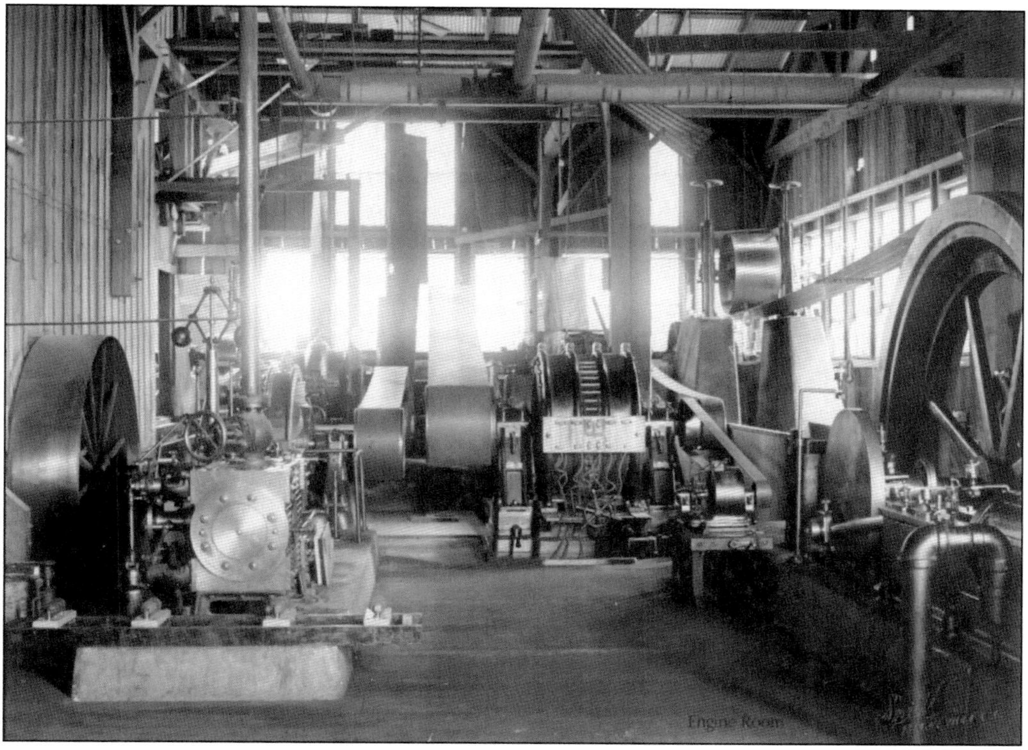

By this time, the refining process had been mechanized and enlarged in scale. Pictured below are the reduction tables. This machine, which washed and separated the gold, was really just an enlargement of more primitive processes. The tables shake, sending the gold to the bottom and allowing the debris to be washed away. Similar tables are still in use today. (Both, courtesy of Ivadene Leech.)

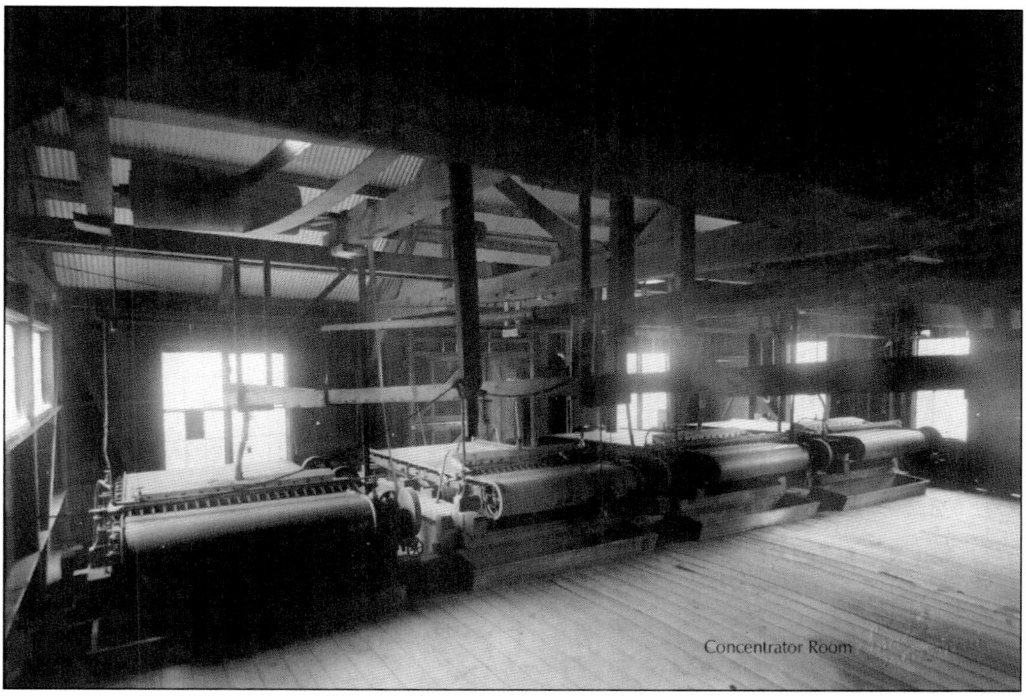

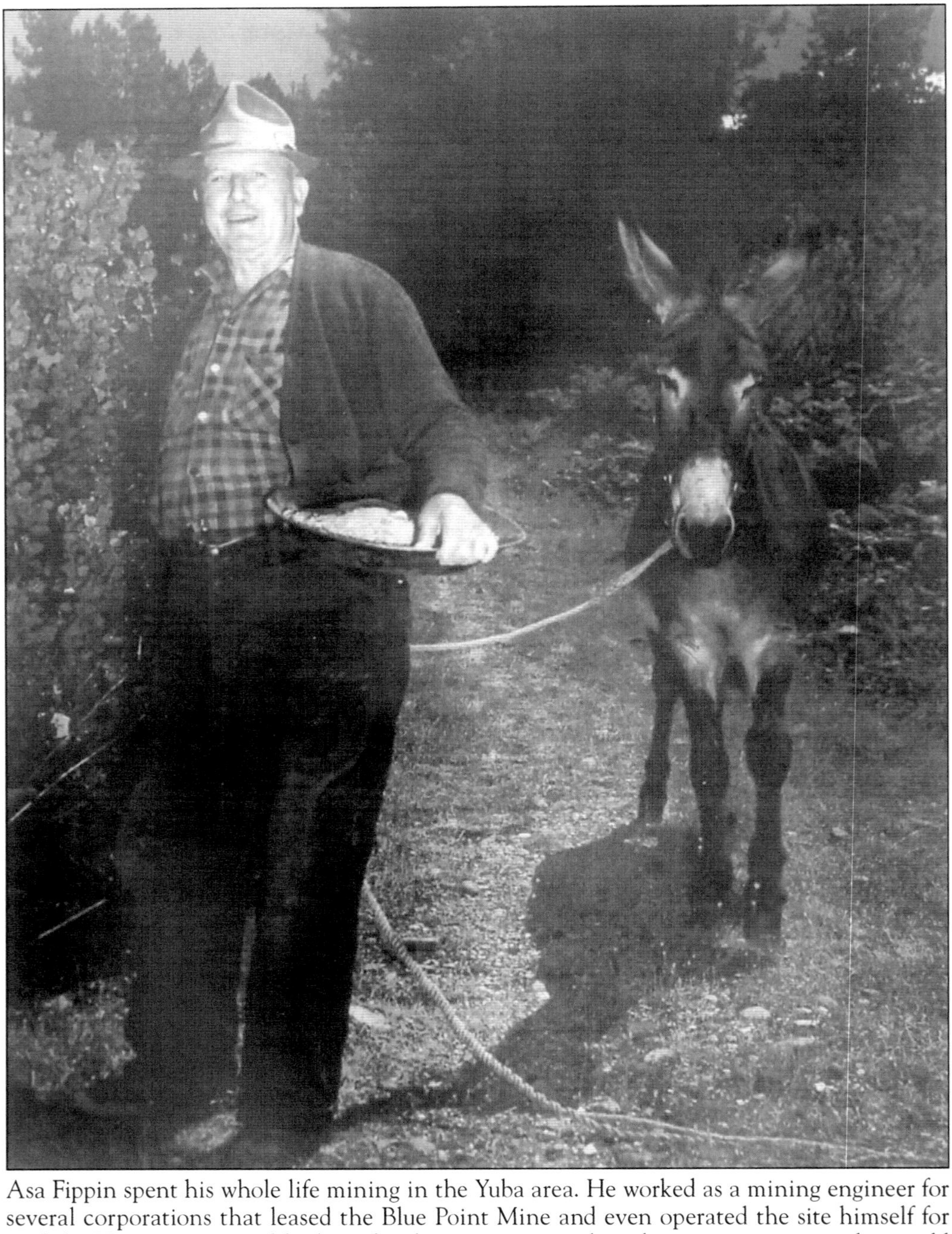

Asa Fippin spent his whole life mining in the Yuba area. He worked as a mining engineer for several corporations that leased the Blue Point Mine and even operated the site himself for a while. He never stopped looking for the precious metal, and even in retirement he would spend an afternoon hiking down to the river with his donkey to pan some gold. (Courtesy of Rosemary Freeland.)

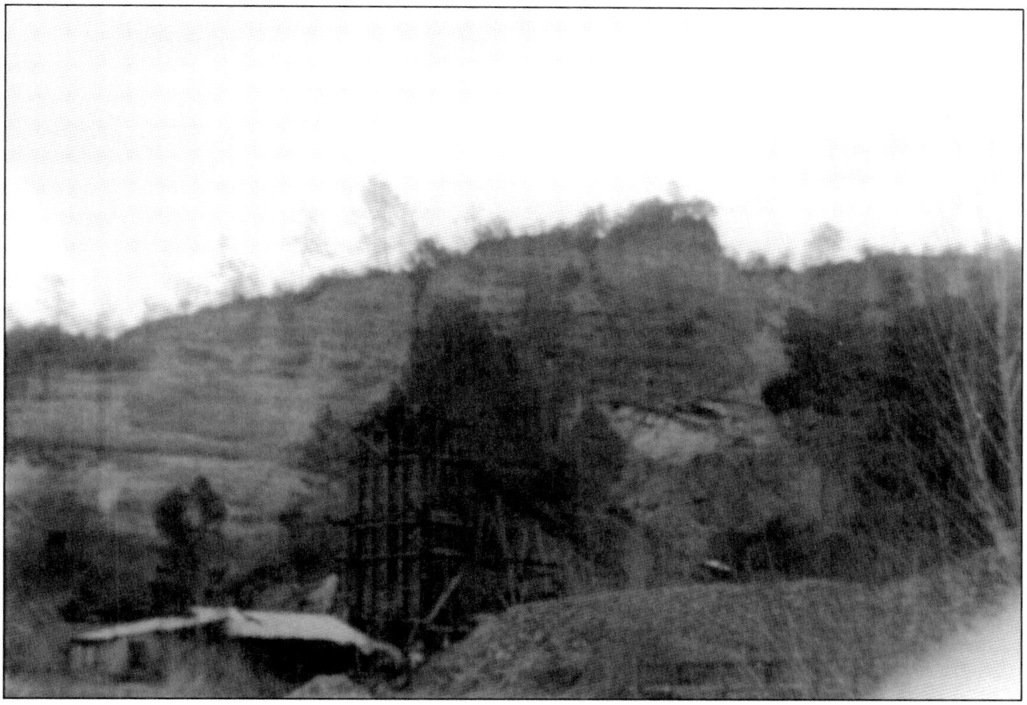

Pictured above is the head frame of Smartsville's Blue Point Mine around 1939. At right, men enter the mine on the Enterprise Bank side. The mine was operated by the Calmich Mining Company at the time, and only produced about eight ounces of gold per day. This is about the same as the output reached by the previous operator, Tintic Standard. Eight men were employed at that time. The mine pit was 360 feet deep and featured some 1,500 feet of tunnels. A drag line was used to draw the gravel to the shaft, where it was loaded into tramcars that took it up to the ore bins, and then put on a trammel. (Both, courtesy of Rosemary Freeland.)

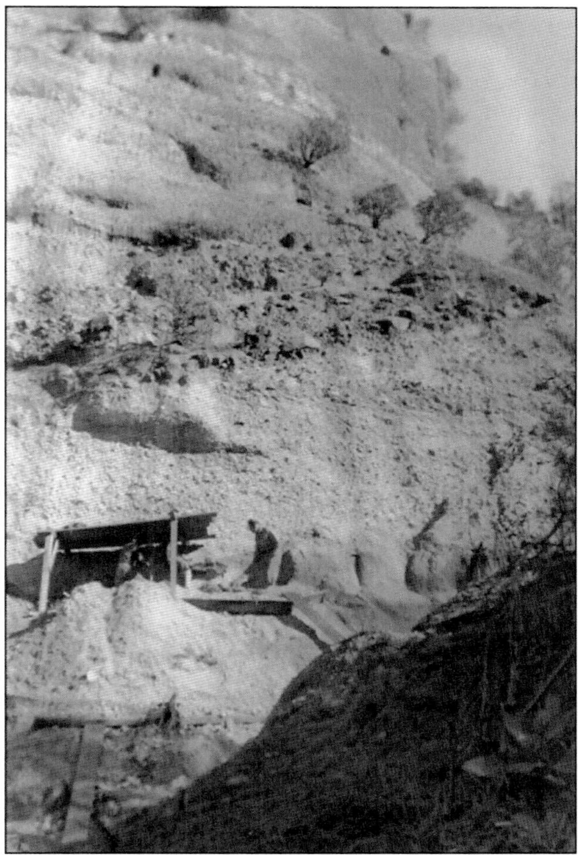

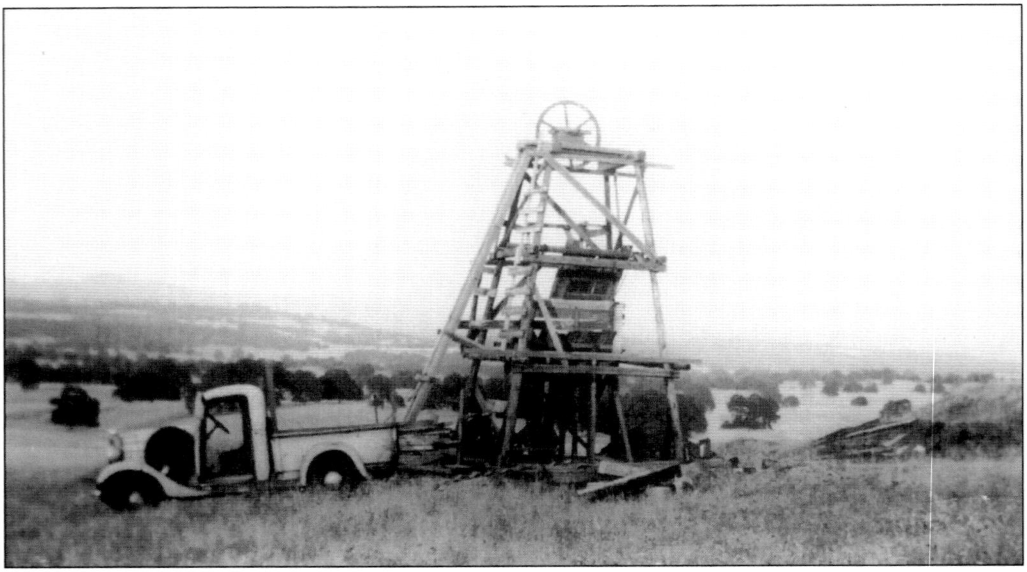

In the early 1940s, Asa Fippin was mining in the Lone Tree area. Below, Sidney and Jessie Fippin—Asa's two sons—pose near the equipment their father was moving into place. The Lone Tree ledge was composed of quartz, so they would have been looking for gold in the rock. (Both, courtesy of Rosemary Freeland.)

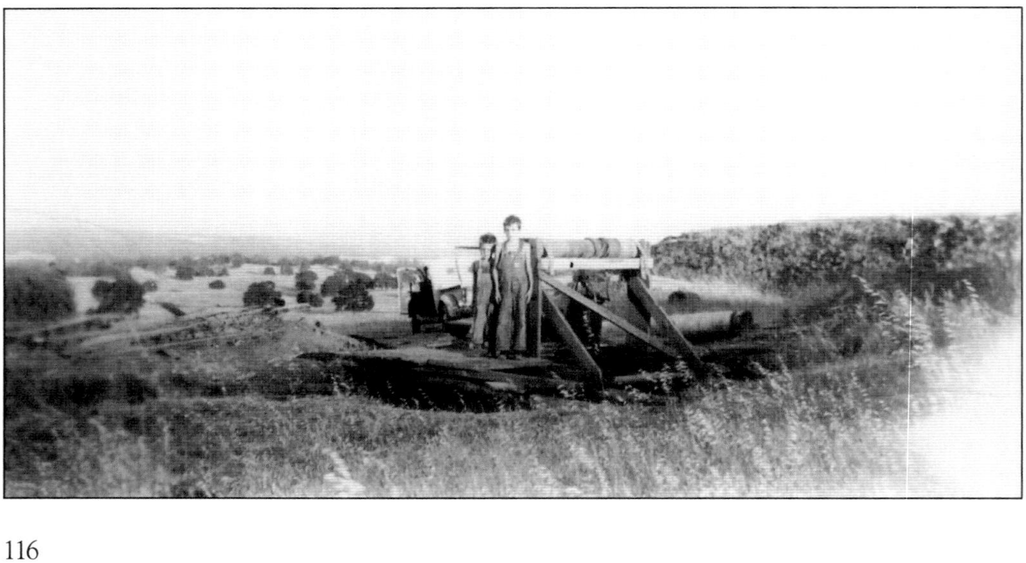

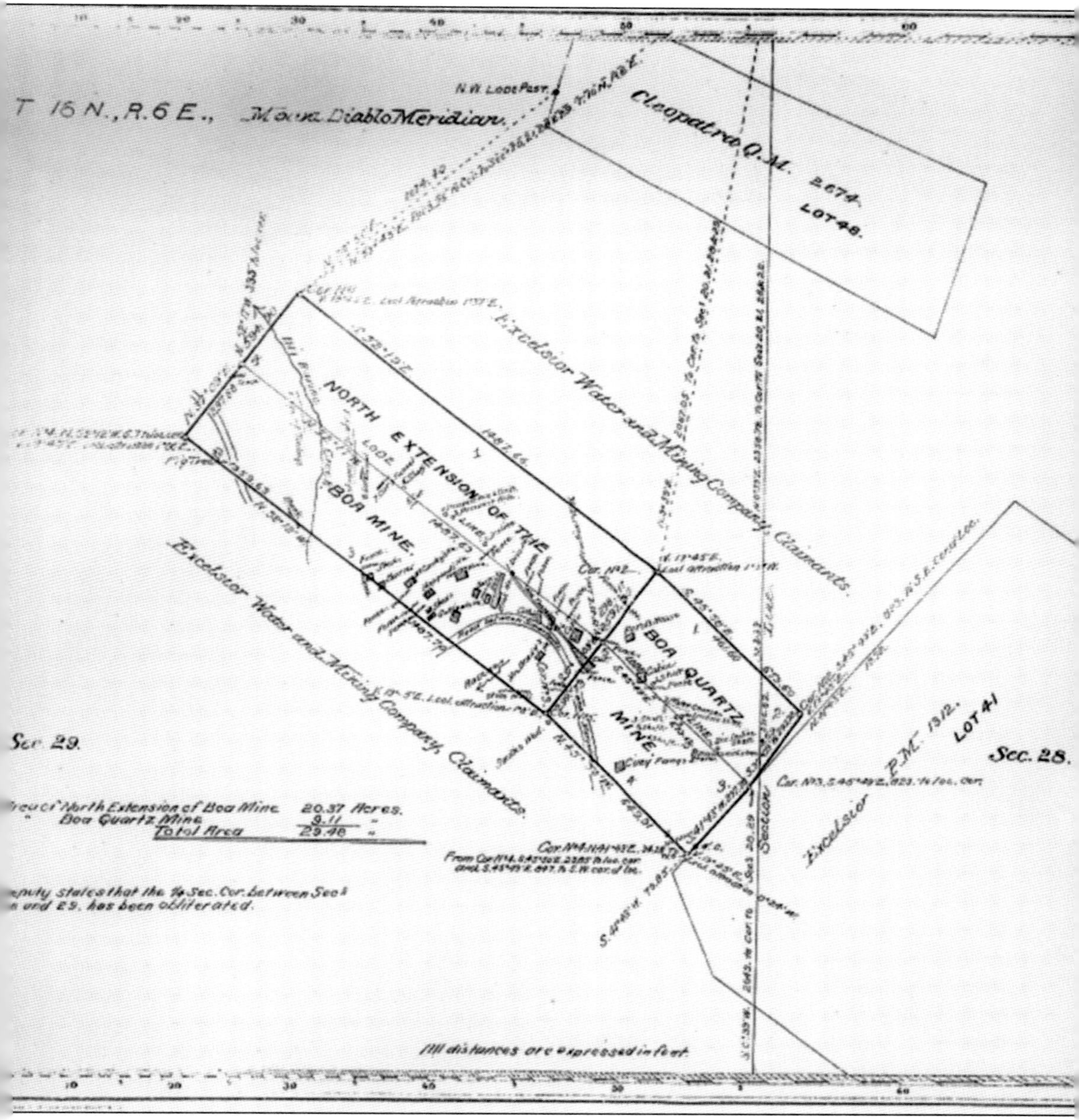

In 1909, a portion of this property was sold to M.C. Meeker, who commissioned a mineral survey to determine if he should try to reopen the Boa Mine and extend the quartz mine there. The mineral survey is full of interesting information, not least of which is the revelation that test shafts had been dug right under where the town stood. On this survey, homes and cabins with people living in them are shown, but the quest for gold seems to have taken precedence over everything else—even in 1909. (Courtesy of the Bureau of Land Management.)

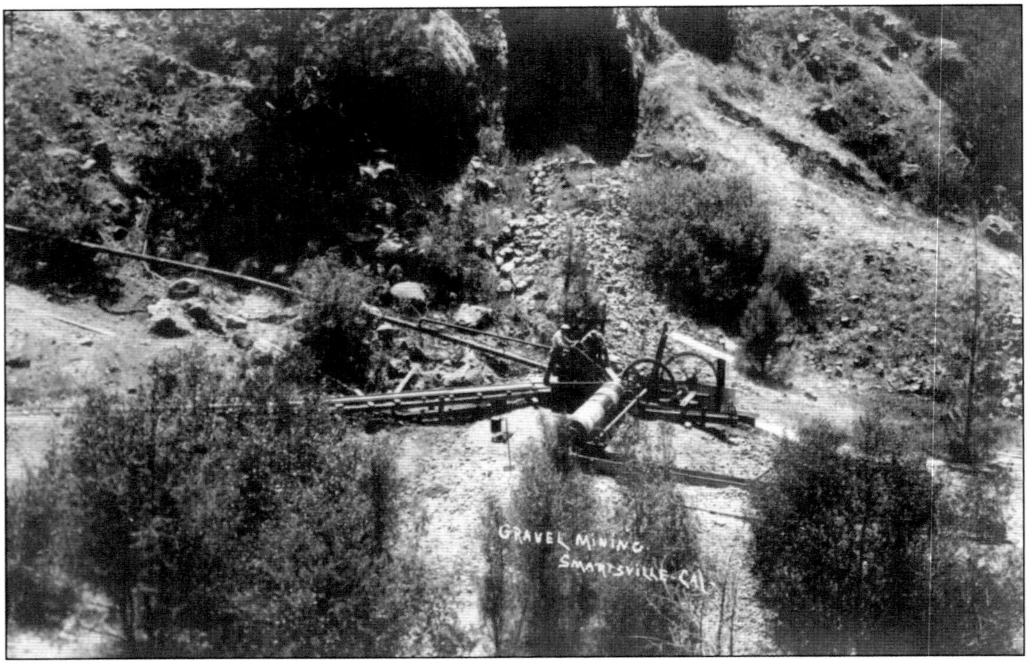

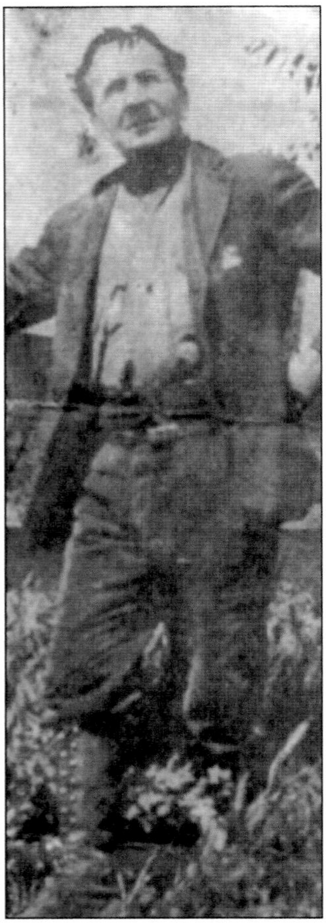

Members of the Hapgood family who still lived in Timbuctoo continued to mine gravel in the area. In this photograph, there appears to be an ore car that runs along a track. They seem to be dumping the material into a mill to process it. (Courtesy of the Marple/Hapgood family.)

In mining terms, a "sniper" is a person who lives a secluded life and pans or washes enough gravel to meet their needs. There were always some snipers living along the Yuba, and they would bring their gold into the store to sell it and purchase supplies. Joe French was one such fellow. (Courtesy of Helen McGovern.)

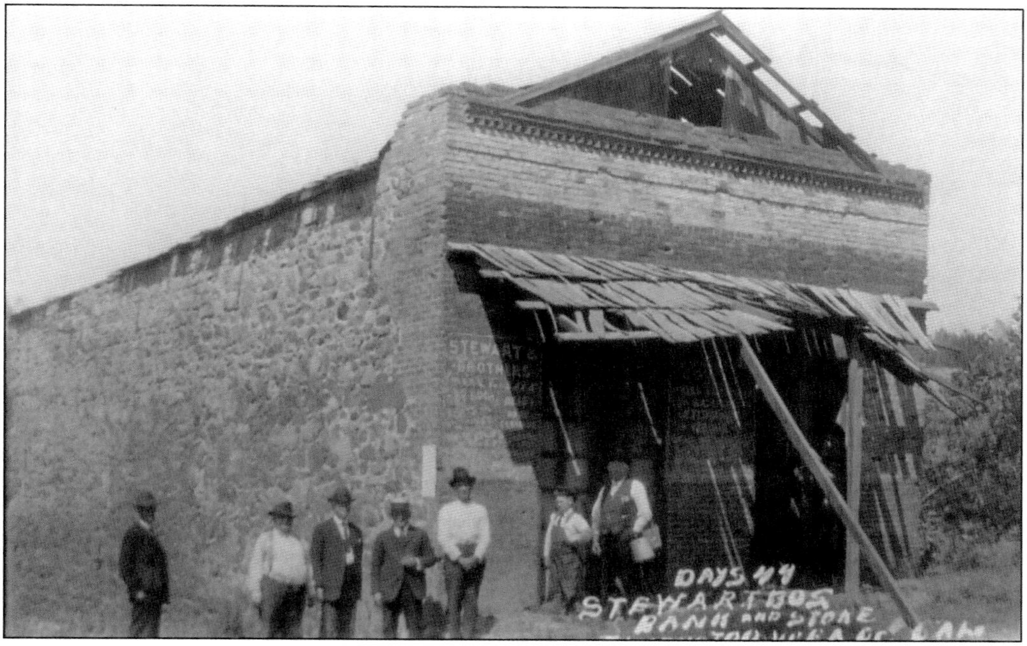

Above, local caretakers show the former Wells Fargo building to a historian writing about Yuba County. The photograph below shows the state historical marker recognizing Timbuctoo as a point of interest. The monument was one-and-a-half miles from Timbuctoo, but after vandals knocked it over, the new marker was placed more than five miles away. (Above, courtesy of Kerrigan/Smith family; below, courtesy of the Yuba County Library.)

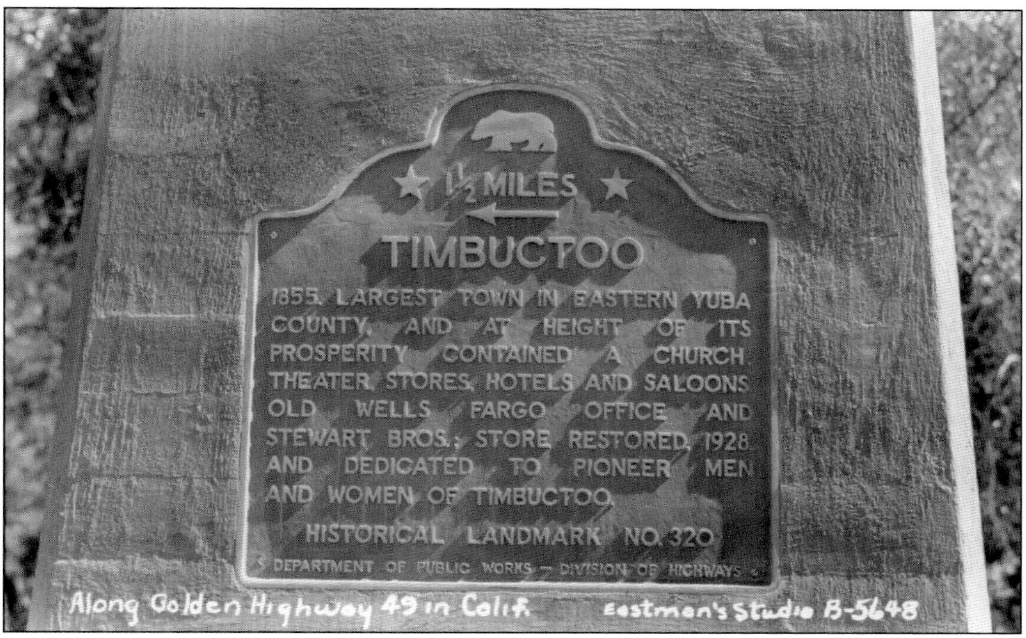

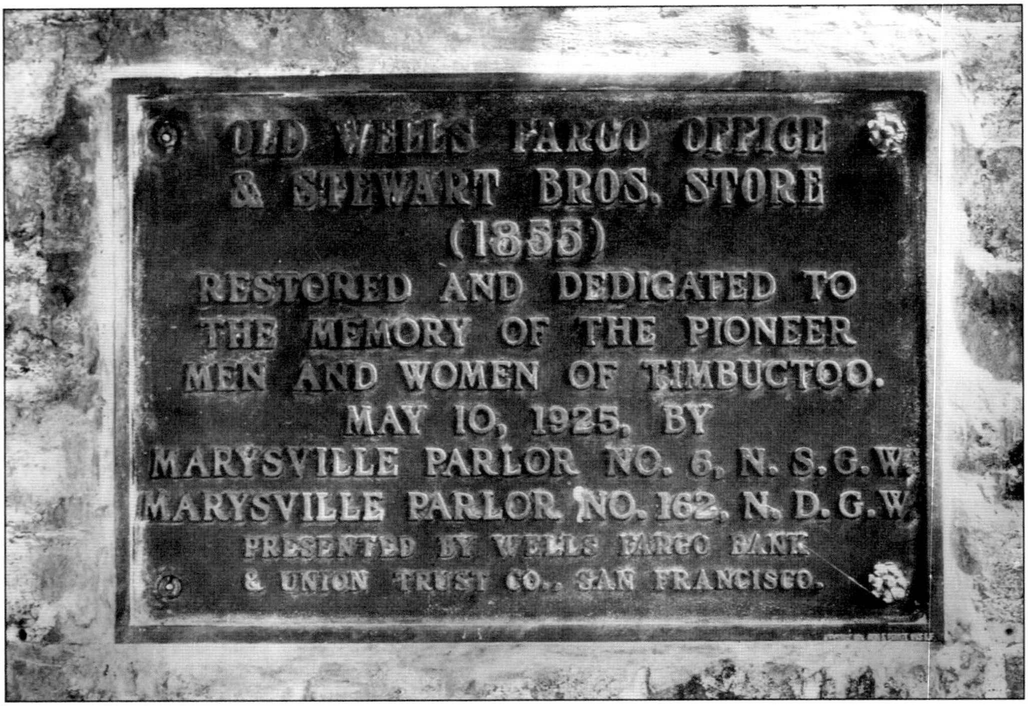

In 1928, the Daughters of the Golden West purchased the former Stewart Bros. Store and Wells Fargo office in Timbuctoo, opening it as a museum. In those days, the site was too far off the main road, and preservation efforts failed due to vandalism and theft. There is hope for these sites, however, as the community is continuing its efforts to preserve local history. (Above, courtesy of Yuba County Library; below, courtesy of Helen McGovern.)

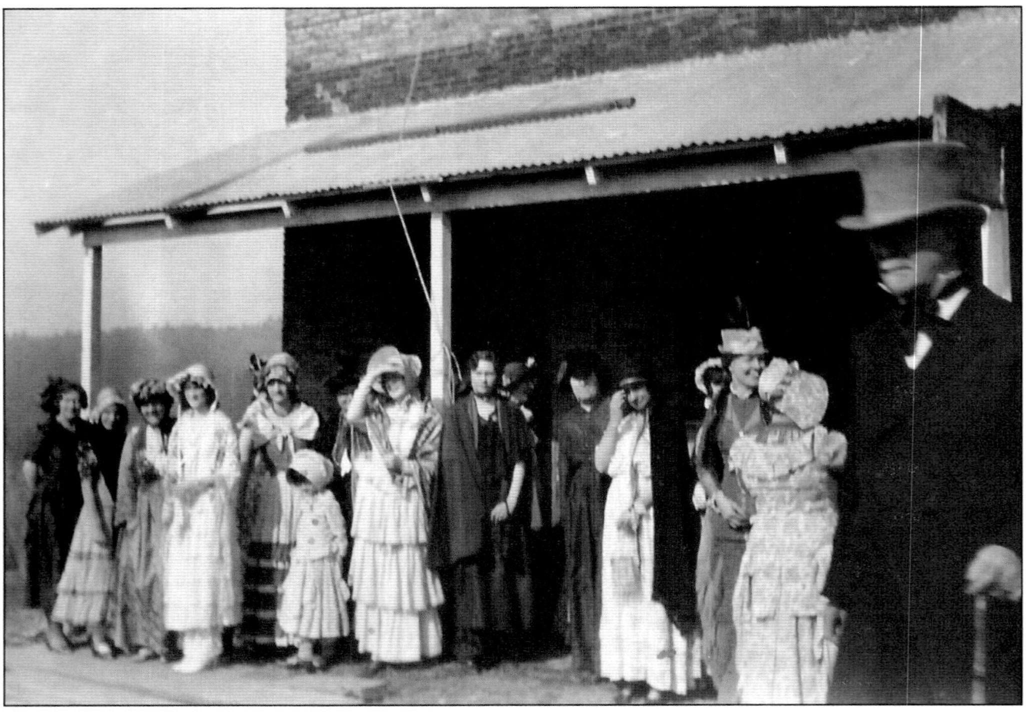

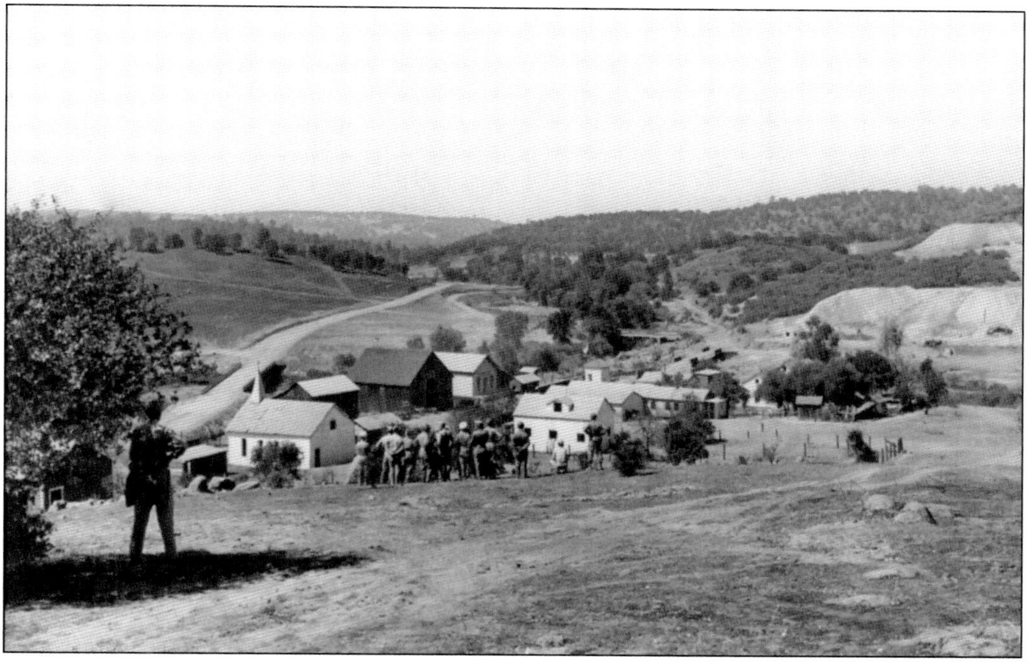

During World War II, the town of Spenceville was appropriated by the military as part of Camp Beale's expansion. The town was turned into a training area called Spenceburg, and the buildings were given German names. Training maneuvers were held there to prepare soldiers from the 13th Armored Division for combat duty in Europe. (Both, courtesy of the US Army Corps of Engineers.)

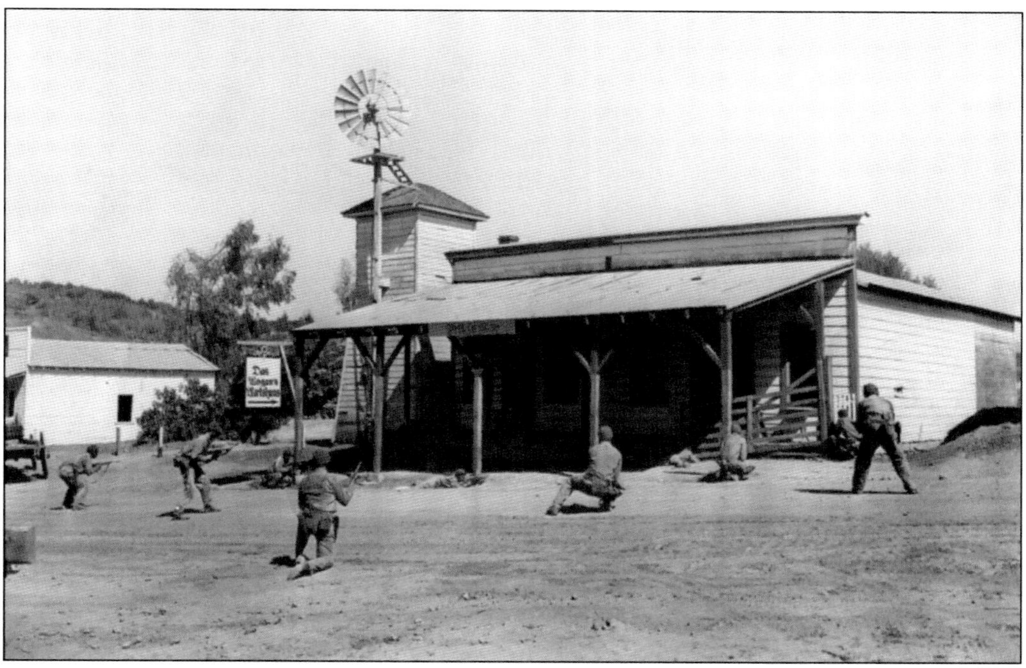

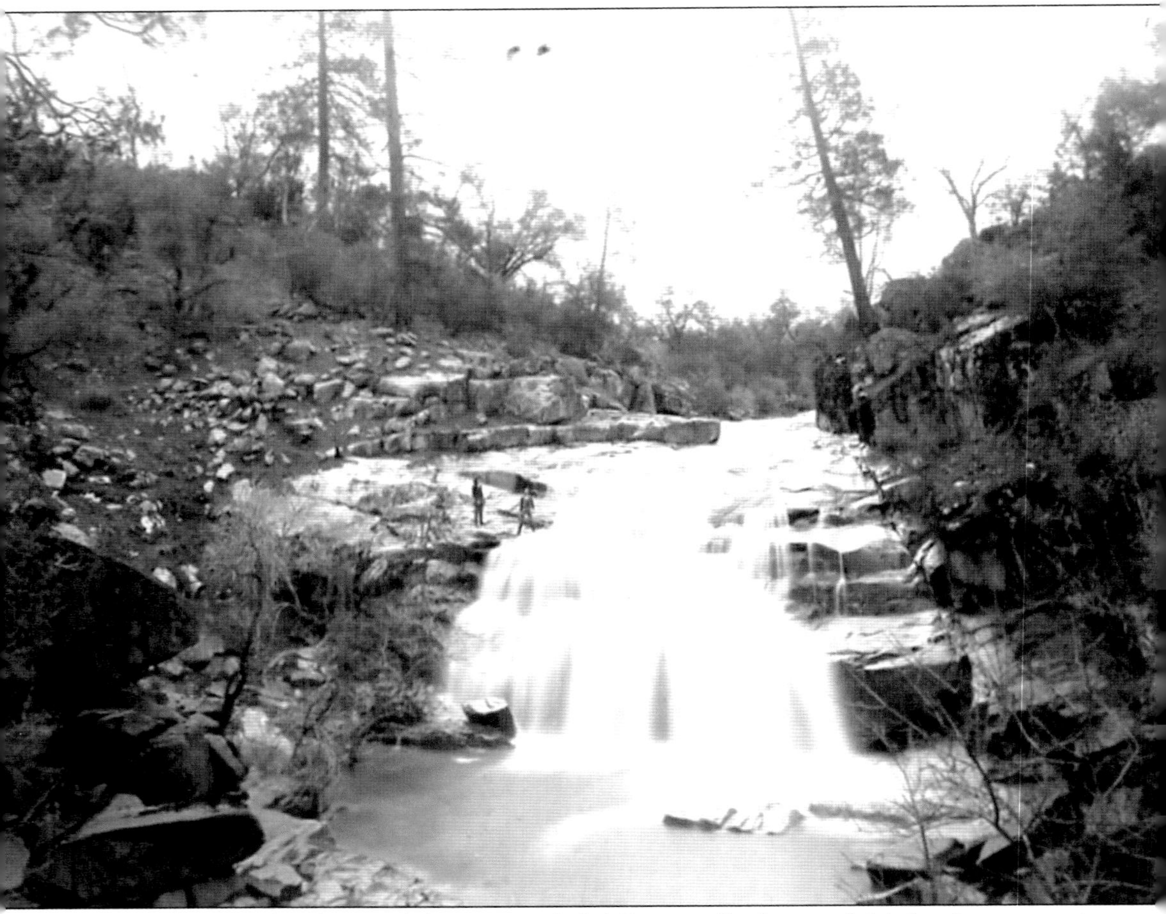

After the mine shut down and the paint works left Spenceville, the mess left behind was staggering. The 10-acre site included 60,000 cubic yards of mine waste and a half-acre, water-filled mine pit. The pit contained 7.5 million gallons of acidic mine drainage that was seeping into two creeks. The flooded pit posed a risk not only to wildlife, but also to people. In 1962, the Spenceville Wildlife Refuge was created and the land was transferred to the Department of Fish and Game. The Spenceville Mine site and old town of Spenceville are within the refuge's boundaries. A massive cleanup of the abandoned mine was undertaken, and the town site, which had been used for training, was cleared of unexploded ordnance. The area is now safe for wildlife and open for recreational use. A beautiful trail allows people to visit the historic Fairy Falls, also called Dry Creek Falls and Shingle Falls.

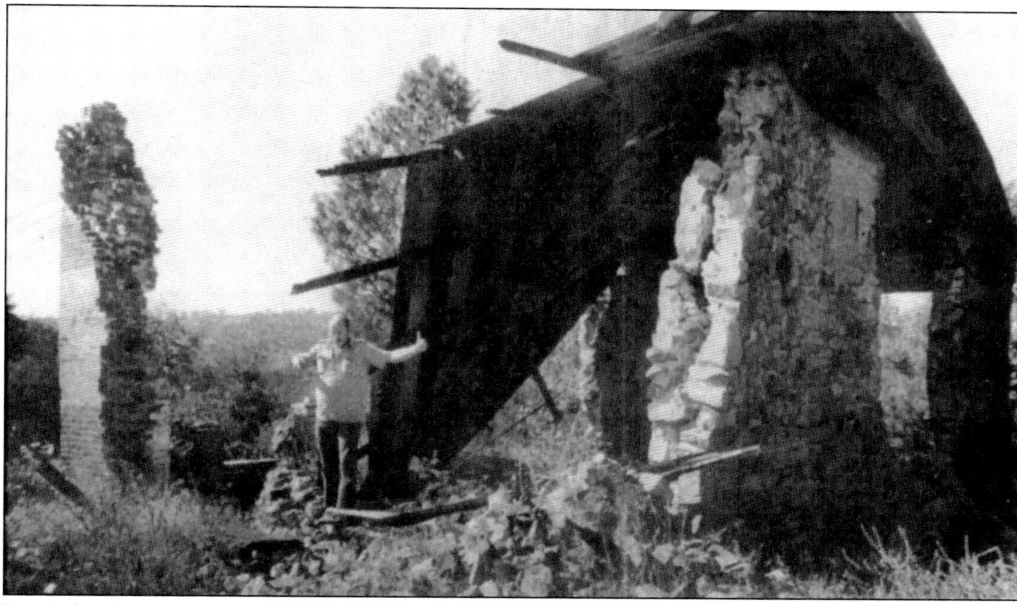

Jim Ashenbrener came to Timbuctoo in the 1960s and decided that he would try to preserve what was left of the town—perhaps even rebuilding the other parts. "Timbuctoo Jim" lived and mined on the property for as long as he could, but his preservation efforts failed. Today, the once-proud Stewart Bros. Store is a heap of bricks with a collapsed metal roof on top. (Courtesy of the Yuba County Library.)

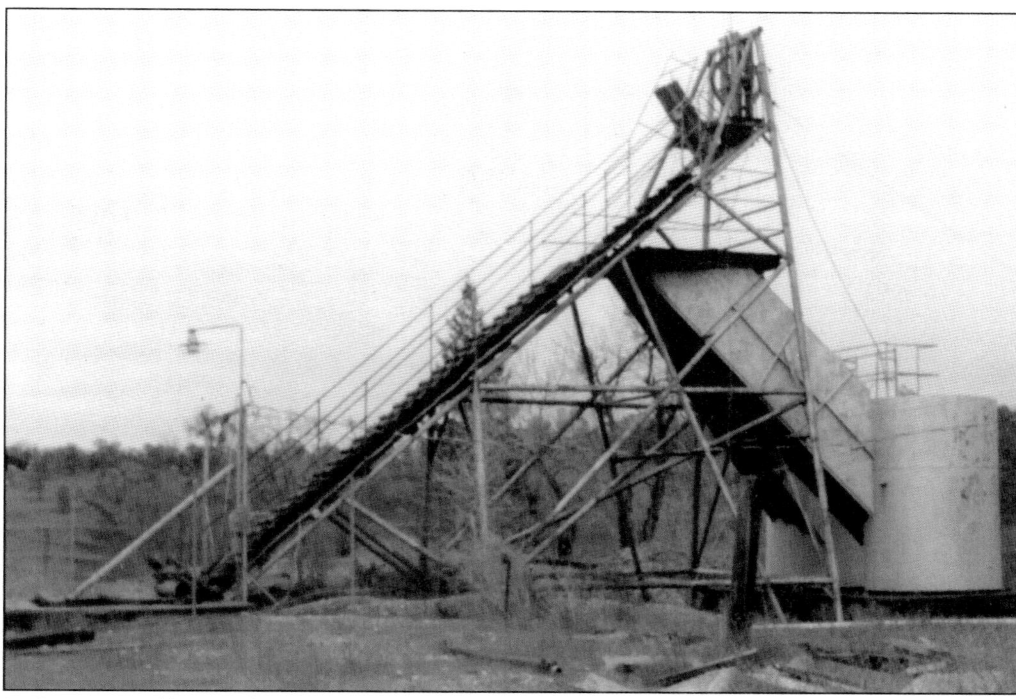

In Brown's Valley, Ruth Mikkelsen and her husband are rebuilding the surface works of the Donnebroge Mine as a historical exhibit. They hold occasional events there, allowing the public to come and see the remnants of the mine and a wonderful collection of mining equipment and information about the area. (Courtesy of Ruth Mikkelsen.)

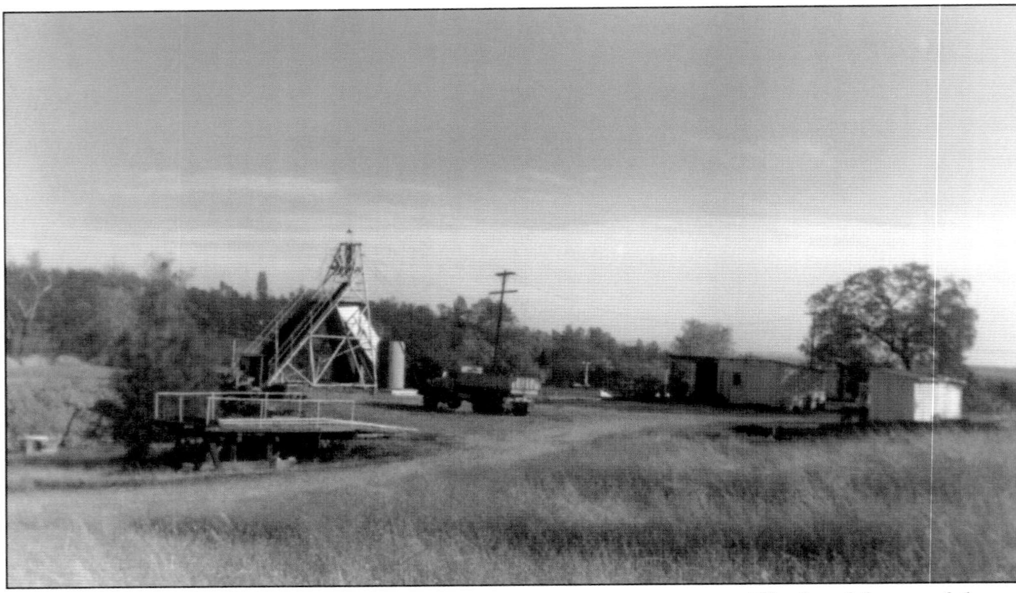

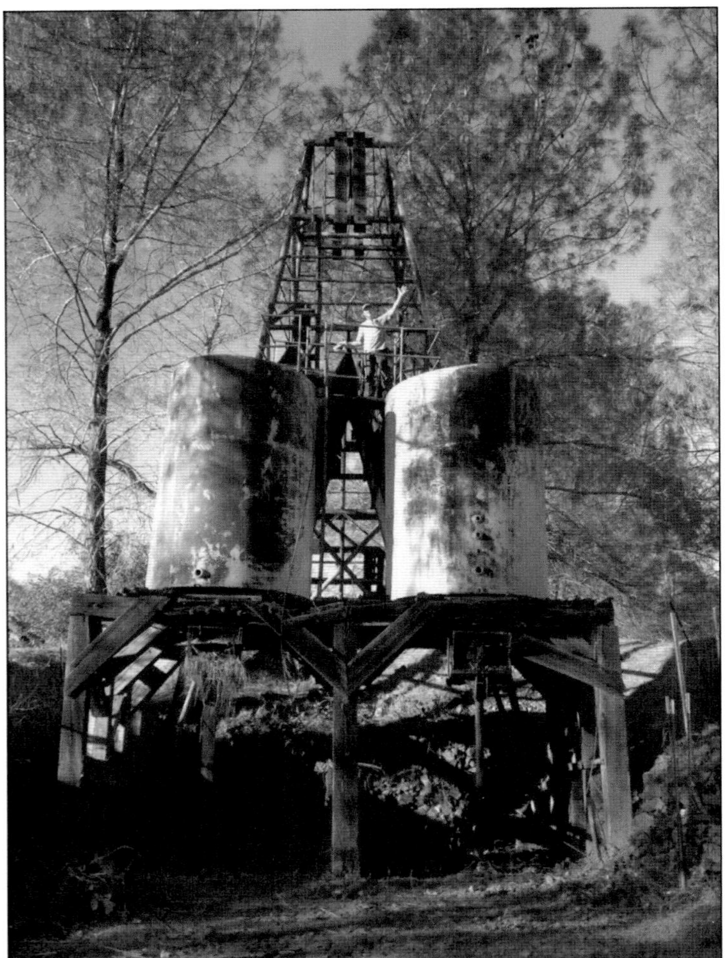

The head frame of the Donnebroge Mine was originally built of wood but it was replaced by a steel unit that has been allowed to decay. Work has commenced to restore the frame to its operational appearance. The ore cars were hauled up the frame and then dumped into the two cylinders below. (Courtesy of Ruth Mikkelsen.)

This photograph, looking up from behind the head frame, shows where trucks would be filled with ore from the cylinders. From there, it would have gone to a processing plant. (Courtesy of Ruth Mikkelsen.)

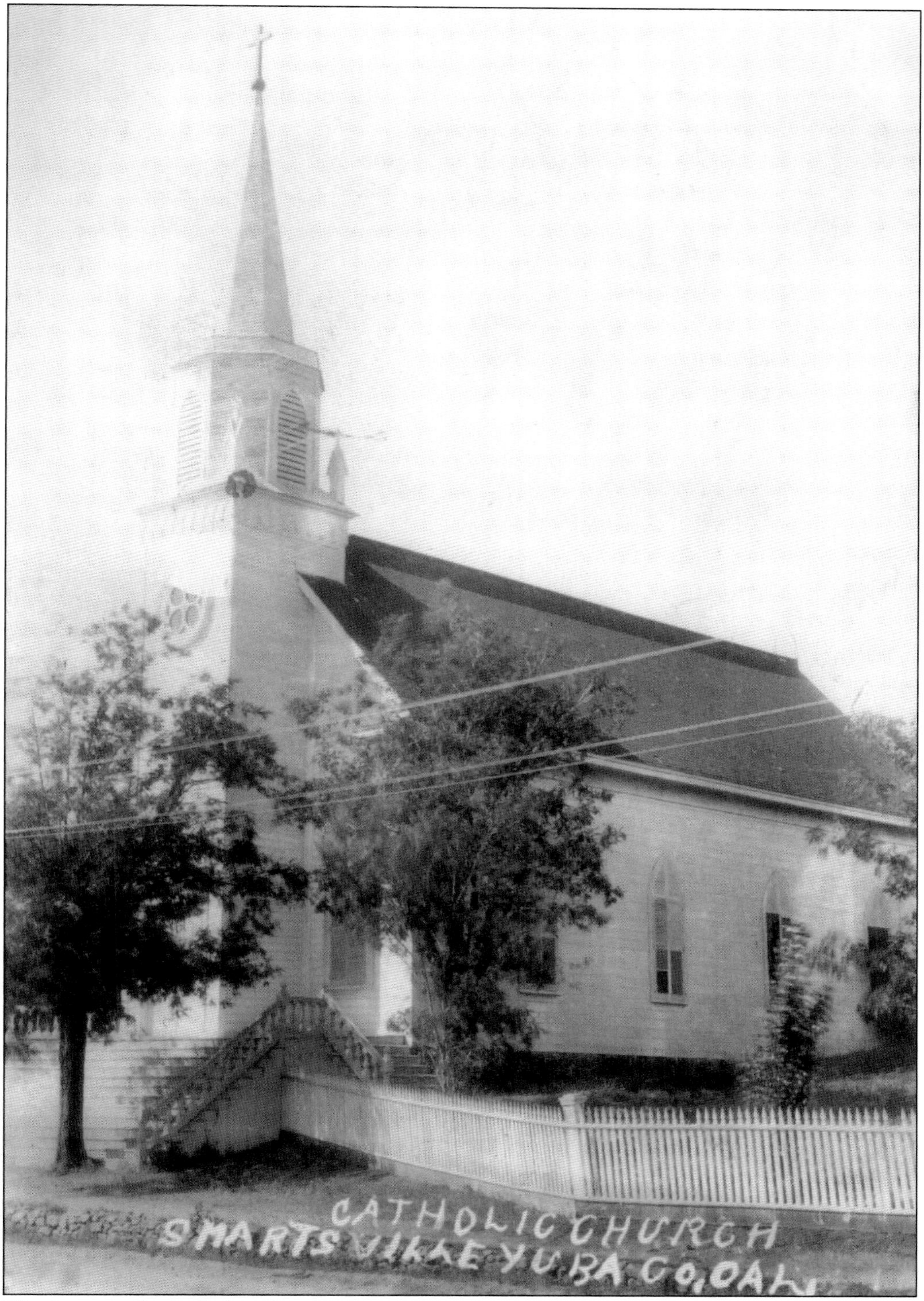

The Smartsville Catholic Church is currently the target of an ongoing preservation effort. In its heyday, the church was the spiritual and communal center of a predominantly Irish immigrant town. It is being restored to be used as a visitor center, host community events, and house a collection of town memorabilia. (Courtesy of the Smartsville Church Restoration Fund Inc.)

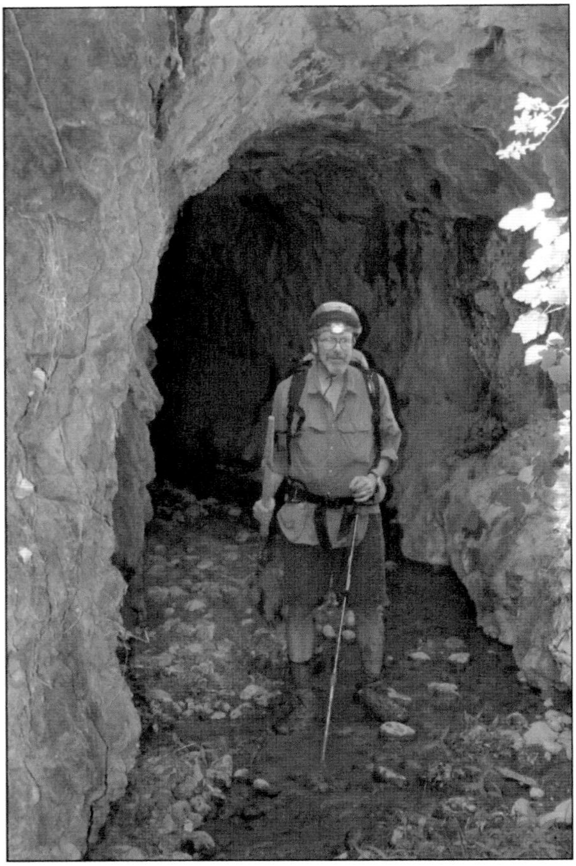

At left, Brian Bisnett stands in the opening of the Blue Point Tunnel. His Excelsior Foundation is the current owner of the Blue Point property and has a vision to preserve the area for public enjoyment. Below, the size of the tunnel is evident as Brian and two researchers begin the 10-mile trek to the other end. George Rigby, the "Sage of Smartsville," wrote in his stories that, "The Blue Point Tunnel is worthy of mention with the seven wonders of the world." (Both, courtesy of the Excelsior Foundation.)

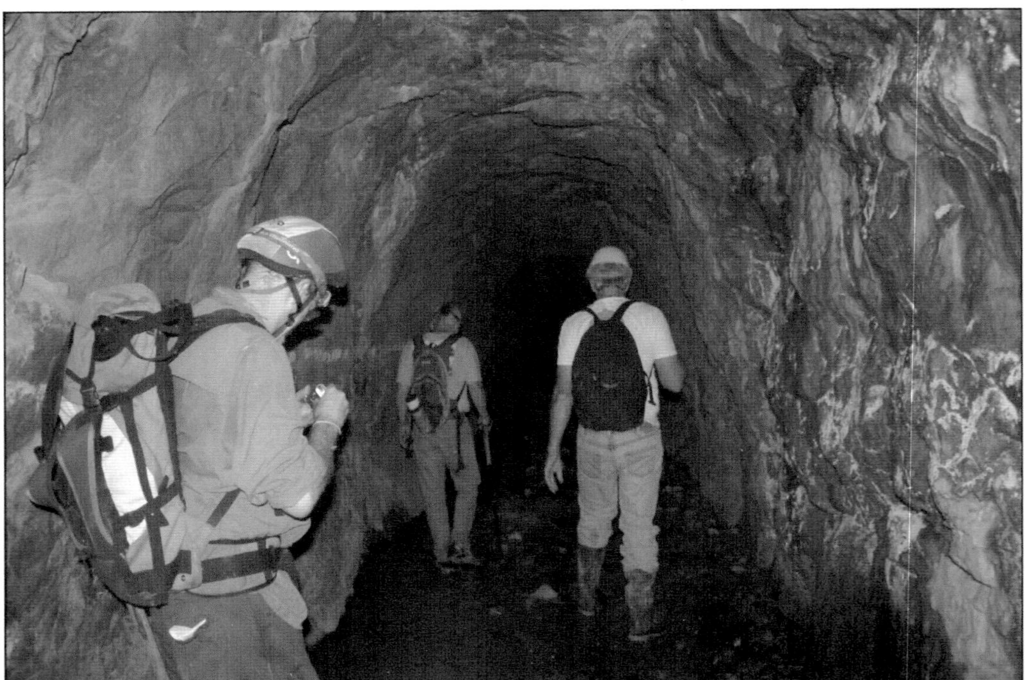

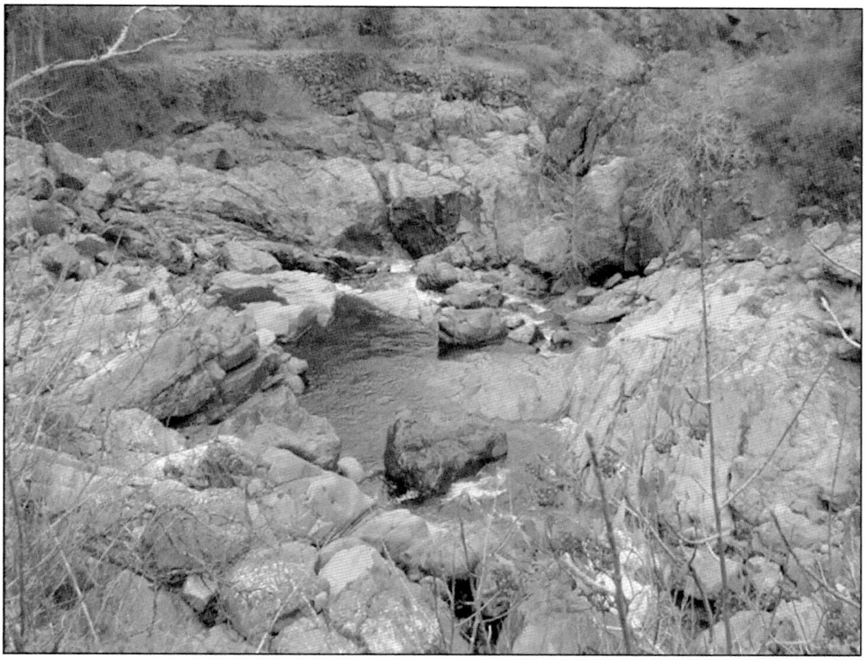

The Excelsior Project is a collaborative conservation effort featuring 924 acres of wetlands, oak woodlands, gold-rush archeological remnants, and miles of critical riparian salmon spawning habitat. The Bear Yuba Land Trust has taken responsibility for one section—now open to the public—called the Black Swan Ponds and Trail. Soon, public access to the Yuba River at Rose's Bar should be available for fishing and paddling. The area that was once nothing more than a place to observe the awful devastation caused by hydraulic mining is now an example of how nature has recovered—providing a place of renewal for local wildlife and human visitors. (Both, courtesy of the Excelsior Foundation.)

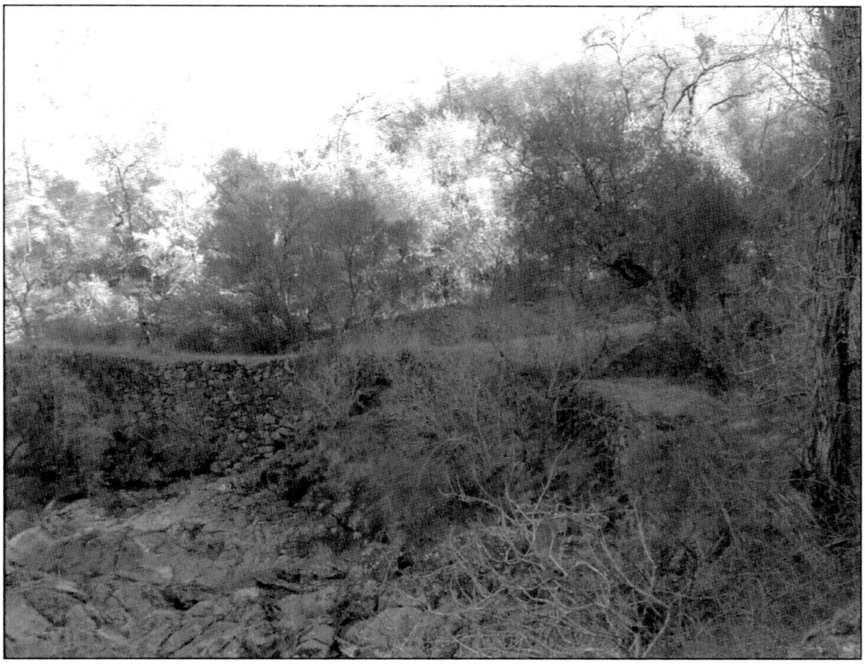